Achieving Supply Chain Integration

Connecting the Supply Chain Inside and Out for Competitive Advantage

Chad W. Autry, PhD
Mark A. Moon, PhD

PearsonPublisher: Paul Boger
Editor-in-Chief: Amy Neidlinger
Executive Editor: Jeanne Glasser
Editorial Assistant: Sandy Fugate
Cover Designer: Alan Clements
Managing Editor: Kristy Hart
Senior Project Editor: Andy Beaster
Copy Editor: Barbara Hacha
Proofreader: Chuck Hutchinson
Indexer: Lisa Stumpf
Senior Compositor: Gloria Schurick
Manufacturing Buyer: Dan Uhrig

© 2016 by Chad W. Autry and Mark A. Moon
Published by Pearson Education, Inc.
Old Tappan, New Jersey 07675

For information about buying this title in bulk quantities, or for special sales opportunities (which may include electronic versions; custom cover designs; and content particular to your business, training goals, marketing focus, or branding interests), please contact our corporate sales department at corpsales@pearsoned.com or (800) 382-3419.

For government sales inquiries, please contact governmentsales@pearsoned.com.

For questions about sales outside the U.S., please contact international@pearsoned.com.

Company and product names mentioned herein are the trademarks or registered trademarks of their respective owners.

Printed in the United States of America

First Printing February 2016

ISBN-10: 0-13-421052-2
ISBN-13: 978-0-13-421052-0

Pearson Education LTD.
Pearson Education Australia PTY, Limited
Pearson Education Singapore, Pte. Ltd.
Pearson Education Asia, Ltd.
Pearson Education Canada, Ltd.
Pearson Educación de Mexico, S.A. de C.V.
Pearson Education—Japan
Pearson Education Malaysia, Pte. Ltd.

Library of Congress Control Number: 2015956647

This book is dedicated to those, both past and present, who have been a part of the family that we call the Department of Marketing and Supply Chain Management at the University of Tennessee, Knoxville's Haslam College of Business. This turns out to be a pretty large family, and it consists of several categories:

- *The faculty.* Throughout the 50+ years that Marketing and Supply Chain Management have resided in the same department at UT, a parade of smart, insightful, and dedicated scholars have passed through and contributed to the ideas articulated here. Particular dedication to Dr. John T. (Tom) Mentzer, who encouraged us to think about integration.

- *The students.* Our undergraduate, master's, and PhD students want to know how they can be successful, as businesspeople or as scholars. They have pushed us to get deeper into the concept of integration and help prepare them to overcome the negative consequences of functional silos.

- *Our industry partners.* At the time of this writing, there are 65 companies who belong to our Supply Chain Forum. They support us financially, but even more important, they talk to us about the challenges they face every day in a dynamic global business environment. Many of those challenges turn out to be challenges surrounding integration, and they have pushed us to find useful, actionable solutions.

So we dedicate this book to this UT family. We also personally dedicate this book to our own families—the Autry and Moon families—who support and love us, and who are ultimately the reason we get up and come to work every day.

Chad Autry and Mark Moon
October 2015
Knoxville, TN

Contents

Preface

It was about 50 years ago, in the mid-1960s, that the dean at what is now the Haslam College of Business at the University of Tennessee decided that he had too many direct reports and that there was too much money being spent on administration in the College. His answer was to take two of the College's academic departments and combine them into a single department. Thus, the Department of Transportation and the Department of Marketing joined together to become the Department of Marketing and Transportation. In the ensuing 50 years, the name of the department morphed into Marketing, Logistics, and Transportation, then again to Marketing and Logistics, and finally to Marketing and Supply Chain Management, which is what we call ourselves today. The organizational restructuring achieved what the Dean hoped—it made the College a little more streamlined and saved some cost in administrator salaries. What he didn't foresee was the strategic "aha" that has evolved over the past half-century in which the faculty of that department have come to embrace the concept of "integration" across functional boundaries.

That strategic "aha" was really inspired by Dr. John T. (Tom) Mentzer, who joined the faculty as Harry J. and Vivienne R. Bruce Chair of Excellence in Business in 1994. Tom Mentzer was a unique scholar in that he was extremely prominent in two separate fields: Marketing and Logistics (later Supply Chain Management). He was at various times in his career the president of *both* the Council for Logistics Management (now the Council for Supply Chain Management Professionals) *and* the Academy of Marketing Science. His larger-than-life personality, and the force of his convictions, helped our department to see the synergy between Marketing and Supply Chain Management, and thanks to his leadership, we developed a vision of business practice that we refer to as *demand and supply integration (DSI)*. Tom helped us get started developing this vision, and since his untimely passing in 2010, those of us who remain at Haslam have continued to develop and refine this DSI vision. We've written articles, both academic and practitioner-oriented, that articulate our thoughts about how demand (sales and marketing) and supply (supply chain) need to be integrated through culture, processes, and tools, for the betterment of the enterprise as a whole.

But we've also come to realize that cross-function integration extends beyond DSI. It includes integration between demand-side functions, such as sales and marketing. It includes integration between supply-side functions, such as procurement and logistics. It includes integration among various supply chain partners in an inter-enterprise context. And it includes a variety of other instances where multiple entities can—and should—behave as a single entity in the pursuit of a common goal. Various UT faculty members, their colleagues at other universities, and their doctoral students have examined elements of integration in the form of articles, books, and even doctoral dissertations. Indeed, "integration" has become a theme of our research over the past several years. As a group, we've uncovered a lot of interesting and useful insights about what integration actually is, how integration can be achieved in complex business organizations, and what the benefits of integration can include.

This focus on integration at the Haslam College of Business, then, is the impetus for this book. We sat back and watched our colleagues do some highly impactful work in this area, and we've done some impactful work ourselves. We came to realize that it was time to bring this work together into a single volume that can help guide practicing business executives through some of the challenges they face in business integration. We approached our departmental colleagues, both on the Marketing and the Supply Chain sides of the department, and asked them to contribute chapters that addressed various aspects of business integration. The result is the 12 chapters found in this volume. These 12 chapters are divided into three subsections: The Foundations of Integration, Internal Integration, and

External Integration. Contributors include faculty from the Department of Marketing and Supply Chain Management at Haslam, faculty from other departments in the College who are also interested in issues surrounding integration, and current and former students who have embraced elements of integration as the focus of their work. We hope that the reader, whether that reader be a practicing business manager who struggles with functional silos and lack of integration, an academic who is interested in pursuing related issues surrounding integration, or a student who would like to learn about the benefits of creating and maintaining an integrative enterprise, will find nuggets of insight and value in these chapters.

So with all that said, let's get into it. We hope you enjoy this exploration through the world of business integration. If you find things you don't like, or don't agree with, or are simply wrong, let us know. Or if you find yourself shouting "Amen!" at anything written here, let us know that too! And Go Vols!

Chad Autry and Mark Moon
October 2015
Knoxville, TN

About the Authors

Dr. Chad W. Autry is the William J. Taylor Professor of Supply Chain Management in the Haslam College of Business at the University of Tennessee. Dr. Autry's professional background includes several years' experience in retail and restaurant operations management. He has worked with and for numerous professional, civic, and governmental organizations related to supply chain process improvement, and has served in leadership positions for the Council of Supply Chain Management Professionals (CSCMP), on the national Board of Directors of the Warehouse Education and Research Council (WERC), and on the local board of the National Association of Purchasing Managers (NAPM).

Dr. Autry's research focuses primarily on socially responsible and collaborative interfirm and interfunctional relationships, their integration within and across firms, and the technological and social issues that support connectivity across multiple organizations simultaneously. He is author of over 70 research studies published in academic and professional outlets including the *Journal of Business Logistics, Journal of Operations Management, Journal of Retailing, Journal of Management, International Journal of Logistics Management*, and *Strategic Management Journal*. He is a co-author of the recent book, *Global Macrotrends and Their Impact on Supply Chain Management*, published by Pearson/Financial Times Press.

Dr. Autry is Editor in Chief of the *Journal of Supply Chain Management* and serves as Associate Editor for the *Journal of Business Logistics, Decision Sciences Journal*, and *Logistique' Management*, in addition to editorial board responsibilities for several other academic and managerial publications.

Dr. Mark A. Moon is an Associate Professor of Marketing at the University of Tennessee's Haslam College of Business and former Head of the Department of Marketing and Supply Chain Management. Prior to joining the Haslam faculty in 1993, Dr. Moon earned his PhD from the University of North Carolina at Chapel Hill. He also holds MBA and BA degrees from the University of Michigan in Ann Arbor. Dr. Moon's professional experience includes positions in sales and marketing with IBM and Xerox. He teaches at the undergraduate, MBA, and Executive MBA levels, and teaches demand planning, forecasting, and marketing strategy in numerous executive programs offered at the Haslam College of Business. Dr. Moon's primary research interests are in Sales and Operations Planning (S&OP), demand forecasting, and buyer/seller relationships. He has published in many of the field's leading journals and conference proceedings. He authored *Demand and Supply Integration: The Key to World Class Demand Forecasting*, and *Sales Forecasting Management: A Demand Management Approach* with Dr. John T. (Tom) Mentzer. His consulting clients have included Honeywell, Goodyear, Corning, Walgreens, Whirlpool, Lockheed-Martin, and many other firms.

1

Integration: What It Is, What It Isn't, and Why You Should Care

**By Mark A. Moon, Chad W. Autry,
and Daniel A. Pellathy**[1]

Prior to the Industrial Revolution, business organizations were far smaller and simpler than they are today. Most businesses, such as family farms, merchant trading rooms, and artisan workshops, were owned and operated by a close-knit group of individuals who sold their goods and services to others within a local community. People specialized in a single trade or craft, exchanged their outputs with others who lived and worked nearby, and sought mainly to provide for family necessities. However, three key innovations that together characterized the Industrial Revolution shifted this model of economic behavior dramatically. The simultaneous emergence of mass manufacturing, long-distance communications, and mechanized transportation threw open the doors to larger and more geographically dispersed consumer markets, providing entrepreneurs who were able to scale up their operations the opportunity to vastly increase their wealth. Business leaders quickly grew their small shops into large, diversified organizations aimed at capturing new demand on national, and later international, markets. The sky was the limit.

To accommodate these revolutionary market shifts, businesses added assets, people, and capital, creating complex multifunctional organizations. Whereas before a few people performed all the tasks required of a business, now entire workgroups were formed to handle the various activities involved with purchasing raw materials, manufacturing and shipping goods, and selling products. The best thinking at the time, exemplified by Adam Smith's widely read treatise on the division of labor, held that organizational performance was maximized by increasing specialization around different activities in the firm. Consistent with this logic, business leaders pushed different functional areas to focus on their particular part of the process, reasoning that by optimizing each set of activities in isolation they could maximize the performance of the organization as a whole. In short, the *era of specialization* was upon us. Moving from a scenario in which everyone did everything to one in which people specialized in different functional activities unleashed massive efficiency gains for early industrial organizations.

The problem is that the times have changed, but the thinking has not. The notion that if each division does its part to the very fullest, the entire organization is sure to succeed still dominates business thinking to this day. We find it in nearly every organization: sales should be the exclusive domain of the sales force; nobody but the accounting group needs to review or understand the financials; manufacturing should worry only about producing finished goods as efficiently as possible. Hire the best talent for each group and focus them on executing just those tasks assigned to their unit. The logic is simple but deeply flawed in today's dynamic market environment.

Research by business scholars in the fields of operations, marketing, and supply chain management points to the conclusion that greater specialization is no longer the engine of growth it once was. Indeed, time and again over the past 30 years, researchers have found tremendous costs associated with the strict specialization paradigm. This "dark side" of overspecialization emerges (a) when activities and priorities in one area become disconnected from activities and priorities in other areas, and (b) when different functional areas lose visibility on the unique value they contribute to their end customer.

The results are wasted resources, internal conflicts, and dissatisfied consumers.

Everyday examples abound. Take, for instance, the all too common practice of salespeople overstating their demand forecasts to ensure product is available for their customers. The result: increased inventories that tie up working capital. Salespeople might be happy, but chances are the enterprise as a whole suffers. Or consider the example of a firm's operations group deciding to source low-cost components halfway around the world. Good for keeping costs down, maybe. But what happens when the firm needs to respond quickly to changes in the marketplace? The operations group may be optimizing on their goal of low unit cost production, but achieving that functional goal may not be in the best interest of the enterprise as a whole. The list of examples goes on. Yet an overemphasis on specialization persists, rooted in people's tendencies to focus on the work at hand and management's tendency to incentivize them on the same. Over time, attending to functional metrics creates the mindset that anything happening outside the business unit is an interference or potential threat. Managers' willingness and ability to cross functional boundaries to maximize organizational performance disappears. And the efficiency gains produced by specialization are quickly outweighed by the loss in effectiveness produced by the disconnect among functions and with customers.

The point is this: if a company is going to succeed in today's dynamic environment, specialization can be only part of the equation. All the parts of the organization that were originally segmented for the sake of efficiency have to be put back together in a way that maximizes customer outcomes and increases profitability. In short, the internal and external functions of a business must become *integrated* for the enterprise to stand a chance. But what does integration entail? The rest of this chapter aims at unpacking this sometimes ambiguous term and pointing the way toward achieving its benefits.

Integration and Supply Chain Management

Integration is at the core of supply chain management. Foundational research in business management had established that optimizing decisions locally within functional areas could—and most likely would—result in suboptimal outcomes for the organization as a whole. Scholars applied this insight in the fields of purchasing, manufacturing operations, and logistics management, spawning what today is recognized as the supply chain field. The centrality of integration is apparent in the earliest definitions of supply chain management, such as the one offered by Oliver and Webber.

> [SCM] views the supply chain as a single entity rather than relegating fragmented responsibilities for various segments in the supply chain to functional areas such as purchasing, manufacturing, distribution, and sales...Supply chain management require[s] a new approach to systems: Integration, not simply interface, is the key.[2]

Likewise, highly influential frameworks offered by Cooper and Mentzer emphasize the importance of integration. Cooper, for example, defined SCM as the "integration of business processes" across key functional areas,[3] whereas Mentzer saw SCM as "the systemic, strategic coordination of the traditional business functions and the tactics across these business functions."[4] More recent reviews of the literature have found that integration both within and across organizations is common to nearly all definitions of supply chain management. This emphasis on integration is also reflected in the practitioner community, where the Council of Supply Chain Management Professionals defines SCM as "an integrating function with primary responsibility for linking major business functions and business processes within and across companies." Investigations into the ways in which integration could be exploited for competitive advantage have also played a significant role in supply chain research. The centrality of integration to supply chain management has even prompted some scholars to suggest it as the field's defining concept.

Given its theoretical and practical importance, it is not surprising that integration has received a great deal of scholarly attention, with the majority of the research seeking to establish its performance benefits. Indeed, empirical evidence gathered over many years suggests that positive associations between integration and various types of business performance do exist. Anecdotal evidence from practitioners has validated these findings over time.

Yet, despite the importance of integration, researchers and practitioners continue to report that companies find it very difficult to achieve. Business practitioners are, if anything, more keenly aware than ever of the benefits of integration, but at the same time, they report that their ability to integrate across key functional areas has not improved meaningfully as knowledge about the subject has grown.[5] This is plausibly due to conceptual issues regarding what scholars and practitioners mean when they use the term "cross-functional integration."

Supply chain management researchers have adopted a variety of perspectives when defining integration and its dimensions. Some researchers have emphasized and studied singular aspects of integration, such as collaboration, interaction/communication, or coordination, whereas others have tried to combine more than one of these terms when conceptualizing integration, such as blending interaction/communication and collaboration, or communication and coordination, within a single concept. Still others have used the term *integration* without specifying integration's "ingredients," creating a catchall phrase that does little to illuminate the more basic concepts that underlie it. As a result, although a key role of scholars within an applied field is to "separate truth from hype," the truth is that scholars have tended to characterize integration in wildly inconsistent ways that are often incompatible with the activities that occur in practice. This lack of a unitary understanding of integration, and the related inability to reliably measure and study it, has served to undermine the best efforts of practitioners and scholars to study the concept or put it into practice. Thus, there remains a compelling need to (1) better define and operationalize the integration concept and (2) advance understanding of the factors that enable companies to successfully develop and maintain integration.

Accordingly, an initial step is to develop a complete understanding of what integration entails. By clearly defining integration and its underlying dimensions, and articulating their relationship to other concepts that serve as antecedents and outcomes, this chapter seeks to provide a solid foundation for scholars and practitioners seeking clarity on this important topic.

What Factors Lead to Integration?

Although rigorous research on integration has not been lacking, the majority of it has focused on the influences of environmental factors that precede or predict integration. Such predictive factors have included environmental variables such as uncertainty, as well as several internal organizational facilitators, such as firm strategies and structures. From the perspective of most managers, however, environmental and/or organizational factors represent institutional constraints rather than decision variables; they impact the firm's ability to integrate, but are largely uncontrollable by managers even in the long run. Thus, the existing research on environmental/organizational antecedents has provided little guidance to managers as to actionable steps that are under their control and that would promote integration in the context of day-to-day operations. This failure on the part of the academy to identify more prescriptive models (to date) has placed managers in the unenviable position of being tasked with developing and maintaining integration with little or no guidance on how to achieve it.

As a result, a young but growing stream of research has begun to focus more intently on how managers can achieve integration across the supply chain by exploring its behavioral antecedents. Behavioral antecedents include the attitudes, behaviors, and decisions exhibited by managers and other employees within the context of day-to-day business operations. Behaviors that appear to enable integration include, for example, demonstrating a cooperative attitude, engaging in informal communication across functional boundaries, gaining an understanding of other functions' activities, and being flexible in decision making. Such behavioral antecedents capture the basic attitudes and actions of individual supply chain professionals and reflect their impacts on the difficult task of achieving integration in a given context. In this sense, behaviors can be thought of as a form of the "soft skills" that have often been identified in the practitioner literature as a critical component for supply chain success, but how they impact the integration of the supply chain remains poorly understood.

What Are Integration's Performance Implications?

The preponderance of the integration literature has sought to establish its performance benefits. Flynn succinctly paraphrased the basic theoretical argument underlying these studies:

> [I]nternal integration recognizes that different departments and functional areas within a firm should operate as part of an integrated process. Because internal integration breaks down functional barriers and engenders cooperation in order to meet the requirements of customers, rather than operating within the functional silos associated with traditional departmentalization and specialization, it is expected to be related to performance.[6]

Scholars have related different forms of integration to improvements in operational effectiveness and efficiency, financial performance (particularly return on assets), successful new product development, customer satisfaction, and market share. Additionally, a growing stream of research also looks at the role of integration in achieving social, environmental, and ethical goals.

Studies on integration have been carried out in a variety of inter-functional contexts. For instance, studies have assessed the integration of purchasing and operations, operations and marketing, and logistics and marketing. These studies have also been carried out across different industries and countries. Positive outcomes discovered across these different contexts bolster the view that integration is indeed linked to performance, and Leuschner[7] and Mackelprang's[8] recent meta-analytic reviews of the empirical evidence support a transcendent linkage.

Solidifying Our Understanding of Integration

Still, despite these seemingly positive results, research on integration remains challenged in two fundamental areas: first, despite the consistency in results, the literature exhibits a stunning lack of cohesion in deriving a consensus definition of integration. This situation has yielded a plethora of operational measures of the concept, which constrains the ability of studies to offer truly generalizable results for practitioners and scholars to rely on. Authors have used several related (yet nevertheless distinct) terms to encapsulate integration, including coordination, collaboration, cooperation, "working together," interaction, and information exchange/dissemination. However, a noticeable lack of attention to the similarities and differences across these terms has led researchers to define and operationalize integration in ways that are generally inconsistent.

For instance, some authors have defined integration in terms of coordinating activities across functional areas, and others have placed greater emphasis on the collaborative efforts needed to maintain common goals toward which activities are directed. Still others have used the terms "coordination" and "collaboration" interchangeably to define integration. Likewise, researchers have used terms such as "information exchange," "information dissemination," and "interaction" to describe integration. Researchers have used these terms generally to cover aspects of both formal information exchange processes and informal communications across functional areas. However, specific definition of variables has ranged from the extent to which information systems are integrated, to whether information is generally shared across functions, to the frequency or amount of communication, to the extent to which there is a common understanding of information. Moreover, there are indications in the literature that at least some of the more formal aspects of basic information exchange, such as having an integrated information management system, may play an antecedent or moderating role in relation to integration rather than constituting one of its dimensions.

The lack of a comprehensive definition of integration and the consequent lack of a reliable operational measure of the concept constrain studies on integration from offering broad-based and generalizable results for both practitioners and scholars. Researchers Frankel and Mollenkopf describe the situation in this way:

> Cross-functional integration (CFI) seems to be one of those notions that we all 'know it when we see it,' but there does not appear to be a consensus about what integration really is... [T]he construct must be clearly defined in order for research results to be meaningfully interpreted across the many streams of literature that include notions of CFI... [A]lthough the concept of CFI has been around for decades, scholars are still in the early stages of genuine construct development.[9]

Toward Consensus on Cross-Functional Integration

Strong conceptual definitions, particularly of established concepts, must be grounded in research. Thus, a comprehensive definition of integration would have to include elements of the multiple perspectives outlined earlier. As such, we propose the following definition of integration:

> Integration is an ongoing process in which functionally diverse areas of an organization collaborate, coordinate, and communicate to arrive at mutually acceptable outcomes for their organization.

According to this definition, integration is conceived as a multidimensional concept that combines elements of collaboration, coordination, and communication. Functional diversification is conceptualized as antecedent to integration; that is, diversification represents the assumed state in which the integration of goals, activities, and knowledge occurs. Likewise, "mutually acceptable outcomes for the organization" are seen as the result of integration, rather than as a dimension of the concept. Definitions of the concept's three dimensions—cross-functional collaboration, cross-functional coordination, and cross-functional communication—are important for completely understanding the phenomenon of integration as it exists in modern business organizations.

Cross-Functional Collaboration

Collaboration generally refers to the mutual establishment of the goals and processes that govern a joint effort. Collaboration represents a special case of the more general concept of *cooperation*. Cooperation can be said to occur in a multi-agent system when (1) agents have a goal in common that no agent could achieve in isolation, (2) agents act to achieve that goal, and (3) agents perform actions that enable or achieve not only their own goals, but also the goals of other agents. Thus, cooperation is centrally concerned with how agents prioritize their own actions with reference to individual and joint goals. It entails not taking advantage of other agents who behave cooperatively. Cooperative agents are therefore willing to make decisions that may suboptimize individual goals in furtherance of a joint goal on the understanding that other agents in the system will behave likewise.

Cooperation incorporates more general ideas found in the integration literature, such as "working together." Note, however, that cooperation does not imply *coordination* (discussed later), insofar as agents can act toward a common goal without any explicit sequencing of decisions or actions. Yan and Dooley make this distinction in arguing that "integration encompasses coordination (alignment of actions) and cooperation (alignment of interests)."[10]

Integration, however, goes beyond simple cooperation to include cross-functional *collaboration*. Collaboration includes working toward common goals, but also entails an ongoing process of establishing those goals and maintaining joint agreement on how best to achieve them. Thus, participants integrate individual goals by negotiating a mutual understanding of group objectives and the role each participant plays in achieving those objectives. In the supply chain context, "collaboration facilitates an assessment of the state of the supply chain, of the needs of the organization, and the determination of an approach for creating and sustaining value based on that collaborative assessment."[11]

Collaboration represents an often difficult process of resolving conflicting interests to establish a joint plan of action with few enforcement mechanisms beyond voluntary agreement. It therefore requires functions to develop meaningful relationships based on trust and mutual respect. It also entails an appreciation of the unique constraints faced by the participants, and may therefore include sharing resources, ideas, and/or information to overcome such constraints. Stank characterizes collaboration in the following way:

Collaboration depends on people's ability to trust each other and to appreciate one another's expertise. It is a voluntary process where two or more departments work together, share resources, and seek to achieve collective goals. It is fundamentally a process that cannot be mandated, programmed, or formalized. Collaboration emphasizes cooperation and is very much 'contingent upon the ability of individuals, scattered within and across organizations to build meaningful relationships.'[12]

At its best, collaboration allows functions to continuously align individual and common goals as they seek to meet the demands of dynamic environments. Based on this understanding, the following definition of cross-functional collaboration is proposed:

Cross-functional collaboration is an ongoing process of jointly defining, adjusting, and working toward common goals while maintaining mutual agreement on how best to achieve them.

Cross-Functional Coordination

Coordination represents a distinct but related concept to collaboration. Whereas collaboration defines common goals, *coordination* refers to the process of bringing together the contributions of constituent members in a way that attempts to consciously optimize a given goal. More colloquially, coordination is determining "what happens when" in achieving some objective. Thus, the central aspect of coordination is the integration of interdependent activities. But, in the context of integration, coordination specifically refers to the process of ordering functional activities—in terms of both substance and timing—so that process inputs and outputs are matched with maximal efficiency. The concept encompasses terms such as "synchronization" and "seamless supply chain operations" insofar as they also relate to inventory control and waste reduction. Germain and Iyer, for example, emphasized coordination in defining integration as the "unified control" of successive supply chain processes aimed at streamlining operations, reducing bullwhip effects, and efficiently matching supply to demand.[13]

Coordination is based on a systems view of the supply chain that sees functional activities as part of ongoing process flows. It may entail, for example, the use of advanced planning systems that employ optimization and metaheuristic approaches to find systemwide solutions or liaison personnel whose specific job it is to coordinate the efforts of several departments. It is important to stress that the concept of coordination presumes a predefined goal. As Oliva and Watson point out: "Coordination...should be considered different from integration in that where coordination takes the target for granted, integration often involves determining this target simultaneously with the aligning of allocation decisions."[14] Based on this understanding, the following definition of cross-functional coordination is proposed:

> *Cross-functional coordination* is an ongoing process of ordering supply chain activities across functional areas based on a systemwide approach that attempts to consciously optimize a given goal.

Cross-Functional Communication

In general, any definition of communication needs to specify (1) what constitutes a communicative act, (2) whether the intention of the sender is considered, and (3) whether the evaluation of the communicative act by the receiver is considered. In the context of integration, communicative acts can take the form of both structured information exchange processes and informal interactions across functions. More importantly, however, the content of these communicative acts represents some tacit and/or explicit knowledge that resides within the sender function. For instance, Mollenkopf refers to information dissemination across marketing and logistics in terms of information regarding products and target customer segments (from marketing to logistics) and warehousing and transportation issues (from logistics to marketing).[15] Other authors have likewise specified the content of communicative acts in terms that indicate the transference of knowledge from one functional area to another. Thus, in the context of integration, a communicative act is not simply the exchange of data or even face-to-face discussions by cross-functional teams; rather, a communicative act is the transfer of knowledge housed in one functional area to other areas of the firm.

Within a supply chain context, moreover, arriving at a shared interpretation of transmitted knowledge is critical to planning and implementing a collective response to the business environment. Thus, the intention of the sender (what the communicative act was meant to communicate) and the evaluation of the receiver (how the communicative act was interpreted) also play an important role in defining cross-functional communication. Indeed, researchers have specifically considered the importance of sender intention and receiver interpretation to integration. More broadly, several research papers have highlighted the need for mutual understanding as a critical element in cross-functional communication. Based on this understanding, the following definition of cross-functional communication is proposed:

> *Cross-functional communication* is an ongoing process of transferring knowledge from one functional area to other areas of the firm so that a mutual understanding of the relevance of the knowledge is achieved.

Extending Previous Definitional Work on Integration

The definitions offered here seek to synthesize previous theoretical work aimed at conceptualizing integration while adding clarity to the terminology employed in the literature. The goal has been to specify the target conceptual domains in a manner that is consistent with prior research. In particular, the definitions offered here clearly build on previous work by Kahn, and Kahn and Mentzer.

In a 1996 article, Kahn provided an influential synthesis of the early literature on cross-functional communication and collaboration:

> Some literature has characterized interdepartmental integration as interaction or communication-related activities, whereas other literature has associated interdepartmental integration with collaboration...There is also a third group of literature, which has implied a multidimensional characterization of integration. This latter perspective conceives interdepartmental integration as subsuming both interaction and collaboration processes.[16]

Building on this later view, Kahn and Mentzer proposed a formal definition of integration as "a process of interdepartmental interaction and interdepartmental collaboration that brings departments together into a cohesive organization."[17] A number of subsequent supply chain management scholars have used this conceptualization as their theoretical basis.

Interaction refers to the set of structured activities between functions that regulate the flow of information between these functions. Kahn operationalized the concept through survey items that ask whether respondents "interact" with other functional areas via meetings, committees, exchange of reports, and so on.[18] Interaction in this sense represents a broad definition of communication that does not specify the content of what is communicated, whether the intention of the sender is considered, or whether the evaluation of the communication by the receiver is considered.

As argued earlier, a more restrictive definition that specifies the transference of operationally relevant knowledge so that mutual understanding is achieved more appropriately captures the underlying concept of cross-functional communication. Indeed, although Kahn operationalizes interaction/communication in broad terms, the author's discussion of the concept suggests a more restrictive understanding:

> Whereas communication should be considered a key component of interdepartmental relationships, viewing integration as 'interaction' prescribes that more meetings and greater information flows should be used to improved product development success. A concern is that more meetings and information flows are not necessarily the answer to improved product development success.[19]

The concern expressed in the preceding passage mirrors the point made by other authors that the central aspect of cross-functional communication is not the exchange of information *per se*, but rather the exchange of operationally relevant functional knowledge. Thus, the definition of cross-functional communication offered in this chapter seeks to build on the concept of interaction established by Kahn but adds specificity in a manner that is consistent with the original conceptualization and the broader literature on integration.

Likewise, our definition of cross-functional collaboration draws on the literature to add specificity to the conceptualization offered by Kahn and Mentzer. Kahn and Mentzer, for example, defined and operationalized collaboration as follows:

> [Collaboration is] an affective and volitional process where departments work together with mutual understanding, common vision, and shared resources to achieve collective goals.

During the past three months, to what degree did your department pursue the following activities with other departments? (Never, Seldom, Occasionally, Often, Quite Frequently)

- Achieve goals collectively
- Have a mutual understanding
- Informally work together
- Share ideas, information, and/or resources
- Share the same vision for the company
- Work together as a team[20]

First, the definition offered here clearly distinguishes collaboration from the more general concept of *cooperation*. This distinction indicates that collaboration goes beyond *achieving* goals collectively (cooperation) to include *defining* goals collectively. Second, "maintaining mutual agreement on priorities in reference to achieving those goals" more clearly specifies the conceptual content of having a "mutual understanding" and "common vision." Third, this notion provides a context for understanding how and why information, ideas, and/or resources might be shared through a collaborative process by focusing attention on the constraints faced by participants. Refining the definition of cross-functional collaboration in these ways is to expect to have implications for its operationalization.

Finally, the definition of integration offered in this chapter adds the dimension of cross-functional coordination to the communication and collaboration elements identified by Kahn. The notion that integration entails the coordination of activities across functions has deep conceptual roots in the supply chain literature. Incorporating this dimension therefore adds an important element to the overall conceptualization of integration.

Planting the Seeds for Integration

It's clear from our interactions with managers and executives that integration is a positive state of being. In managerial practice, programs such as sales and operations planning (S&OP) have been implemented at hundreds of companies in an effort to achieve this integration, yet many of those companies would not describe themselves as truly integrated. The question remains, under what conditions can a company achieve this worthy goal, especially in a complex, potentially global enterprise? We propose that three conditions exist that provide the best environment for business integration to flourish: organizational structure, process, and culture. To illustrate our examples, we use S&OP as the specific context for presenting our ideas about the ideal conditions for fostering organizational integration. However, we would expect the same conditions to exist in many other integration-related settings as well.

Organizational Structure

By organizational structure, we refer to the reporting relationships that exist in a firm. In the internal supply chain of a company, organizational structure can encourage integration if a process is organizationally aligned with other functions of the enterprise. The most valuable integration opportunities tend to come when a function that is "upstream facing" integration with others that are "downstream facing," that is, when operations or logistics integrate with sales or marketing. However, these types of integration often present the biggest challenges. Such is the case of S&OP, which is often perceived, at least by sales and marketing people, as "supply chain planning" when it should be perceived as integrated *business* planning. By organizationally aligning the S&OP process with sales or marketing, such misperception can be addressed. Similar considerations can be made in the case of the organizational "home" of the forecasting, or demand planning, function in a firm. Many firms house demand planning in the supply chain group, for reasons such as "we don't trust sales and marketing to prepare accurate forecasts." Organizationally aligning demand planning in the sales or marketing group, where demand actually occurs, can potentially contribute to integration.

One way that companies often use organizational structure in an attempt to drive integration is through a matrix organizational structure. For example, an S&OP process owner could find him or herself in a matrixed role, reporting to both a sales leader and a supply chain leader simultaneously. Or, in a variation on that theme, that individual could be "solid line" to the sales leader and "dotted line" to a supply chain leader. Although simple to execute, such a strategy often creates the illusion of integration, rather than true integration. Although such matrix approaches can encourage individuals to be cognizant of the needs of multiple functions, it can also lead to significant role conflict or role ambiguity in the individuals involved. Without attention to the other two drivers of integration— integrative processes and a culture that facilitates integration—such organizational structure strategies are unlikely to lead to true integration.

Process

Processes are formal, disciplined mechanisms that bring together relevant pieces of information, from different points of view, delivered by different people, in a regularly scheduled forum, to help the organization make decisions that will help it achieve its goals. Such processes are referred to as S&OP, SIOP, IBP, DSI, or other labels. A good example is the well-documented integrated business planning process that is normally associated with the consulting firm Oliver Wight. From a high-level view, it typically consists of five separate steps: Product and Portfolio Planning, Demand Planning, Supply Planning, Financial Reconciliation, and Executive Review. Each step is often documented with detailed flowcharts that describe the sequence of events that must occur, the analyses that need to be completed, and the timing of those analyses. Such a process is often repeated on a regular, monthly drumbeat. Information is brought together from multiple functions in the firm, including sales, marketing, supply chain, finance, and senior management. Customers and suppliers are often represented in the different stages of the process. Companies frequently spend large amounts of time and effort to construct and document these processes, and they are often elegantly designed and comprehensive. Unfortunately, it is our contention that these processes, by themselves, often fail to achieve true integration. Both organizational structure and integrative processes are necessary but not sufficient to the goal of true integration. The final mechanism, culture, must be addressed.

Culture

Defining culture is difficult. John Mello has published articles in this and other outlets in which he has commented upon the effect that culture has on effective forecasting and business integration.[21] Merriam-Webster defines culture as "a way of thinking, behaving, or working that exists in a place or organization (such as a business)." An organization's culture can be observed in the norms of behavior and attitude that are present in a firm. How people think; how they interact with others; what they find important; how hard they work; how they dress—all these and countless others define an organization's culture.

Some organizational cultures are supportive of integration, and some are resistant. Those that are resistant to integration are characterized by each functional group having its own unique culture, and the people are distrustful, or even disdainful, of other functional groups' cultures. In a business integration context, this can be manifested in the following types of statements:

"I don't believe any of the forecasts coming out of sales. They're way too optimistic."

"All the supply chain people care about is minimizing inventory. They don't care about serving our customers."

"Finance is living in dreamland. We'll never make that Annual Operating Plan number."

On the other hand, a culture that promotes integration is one where people are pursuing common goals, regardless of the functional area in which they work. So what can a company do to create that integration-friendly culture? Or, what can a company do to transform an integration-*unfriendly* culture into one where integration can thrive? It is our assertion that there are two approaches to addressing these problems, *both of which must be addressed*: top-down and bottom-up.

Tools Available to Managers

Top-Down Culture Change

The signals that people receive from those who are above them in an organization influence their behaviors and their attitudes. This means that enterprise leaders must send very clear, consistent signals that integration is a business imperative and that everyone must behave in this way. In any enterprise, the C-suite executives—CEO, Chief Demand Officer (whether that be the head of sales, the head of marketing, or both), Chief Supply Officer (which might be a combination of head of supply chain and head of manufacturing), and the CFO—must say, and more importantly do, everything possible to communicate that integrative behavior is expected.

The most important piece of top-down culture change is *what senior leaders do*, not *what senior leaders say* (although what they say is important, too). They have to be willing to expend resources to get the right tools and people in place to support the integrative business processes. They have to be willing to look at measurement and incentive systems that are in place, to be sure that integrative behaviors are in fact rewarded. And they have to model those behaviors; they have to regularly attend and engage in the executive S&OP meetings and show willingness to sometimes sacrifice their own functional objectives to reach common objectives.

Importantly, the one individual that *must* play this leadership role is the Chief Demand Officer. Consistent with our previous comments, one of the most common causes of S&OP failure is lack of engagement from the demand side of the enterprise—sales, marketing, or both. Several companies have described their S&OP processes as being "&OP—sales is nowhere to be found." The Chief Supply Officer is usually the driver of these integrative processes, so he or she is usually a believer. So the greatest challenge to creating this top-down culture change is to convince both the CEO and the Chief Demand Officer that these integrative processes must be put in place and supported with committed behaviors from those involved.

Bottom-Up Culture Change

Although the impetus for the culture change needed to achieve true business integration must start at the top of the organization, integration is unlikely to occur just because the CEO wants it to happen. So what can be done to drive these integrative behaviors on the part of the people actually doing the work? Focus should be placed in two areas: incentive and measurement strategies and education and training.

A useful piece of folk wisdom can be found in the phrase "what gets measured gets rewarded, and what gets rewarded gets done."

In this context, this folk wisdom suggests that if you want individuals to engage in integrative behaviors, encourage such behaviors though their compensation structures or their performance plans. For example, most organizations benefit from receiving demand-forecasting input from their sales teams. This would be an example of a valuable integrative behavior. However, in many companies this behavior is neither measured nor rewarded. Without measuring this contribution, and acknowledging that contribution in either the compensation structure or individual performance plans, it is not surprising if salespeople either spend very little time on the task, or even worse, if they intentionally provide *bad* information in order to advance a different agenda. Thus, the measurement and reward strategy can incentivize integrative behaviors. So bottom-up cultural change can be initiated and reinforced by closely examining the way all people are measured and rewarded. Senior leaders need to look carefully at what drives individual decision making, and finding ways to measure and reward integrative action must be a priority.

The second way that culture change can take place is through education and training. The training that is most impactful for driving organizational change is when individuals from multiple functional silos sit in a classroom together to learn about the benefits of integration, and how they can individually contribute to that integration. Many times, "aha" moments take place when individuals from sales first hear what happens to the forecasts that they submit. "I had no idea that my forecast had that impact," they say. "I thought I was just gaming my future quota numbers. You mean you actually take that forecast and make supply chain decisions based on those numbers? Are you kidding me?" Extremely useful classroom experiences can occur when people from sales, marketing, logistics, procurement, operations, finance, and demand planning are all in the same training class. One useful mechanism is to run a simulation and assign salespeople to logistics roles, or procurement people to marketing roles, or finance people to sales roles. Real moments of insight occur when people experience the effects that their nonintegrative behaviors have on the company.

Clearly, bottom-up culture change must be planned and managed. It doesn't happen on its own.

Conclusion

In summary, then, important points to remember from this chapter are the following:

- S&OP, or other similarly named processes, often fail to achieve true integration.
- Integration should be thought of as multiple entities *behaving as if they were a single entity* to achieve *common organizational goals*.
- Integration can be achieved through multiple mechanisms: organizational structure, integrative processes, and organizational culture. Culture is, by far, the most important, yet the most difficult, to put into effect.
- Efforts to achieve true business integration must be driven both from the top down and from the bottom up. Top-down change is driven by senior leadership commitment to an organizational structure that will not impede integration, formal disciplined processes that create a forum for integration, and a culture that will facilitate integration. Bottom-up efforts to achieve integration should be driven by measurement and reward structures that incentivize integrative behaviors, and education and training opportunities that demonstrate to individual people the benefits that can derive from true business integration.

Endnotes

1. Mark A. Moon is the Department Head of Marketing and Supply Chain Management at the University of Tennessee's Haslam College of Business. Chad W. Autry is the W.J. Taylor Professor of Supply Chain Management, and Daniel J. Pellathy is a Supply Chain Management doctoral candidate in the same department.

2. Oliver, R.K., and M.D. Webber. 1982. Supply-Chain Management: Logistics Catches Up with Strategy. *Outlook* 5(1): 42–47.

3. Cooper, M.C., D.M. Lambert, and J.D. Pagh. 1997. Supply Chain Management: More Than a New Name for Logistics. *The International Journal of Logistics Management* 8(1): 1–14.

4. Mentzer, J.T., et al. 2001. Defining Supply Chain Management. *Journal of Business Logistics* 22(2): 1–25.

5. "Henry" Jin, Y., A.M. Fawcett, and S.E. Fawcett. 2013. Awareness Is Not Enough: Commitment and Performance Implications of Supply Chain Integration. *International Journal of Physical Distribution & Logistics Management* 43(3): 205–230.

6. Flynn, B.B., B. Huo, and X. Zhao. 2010. The Impact of Supply Chain Integration on Performance: A Contingency and Configuration Approach. *Journal of Operations Management* 28(1): 58–71.

7. Leuschner, R., D.S. Rogers, and F.F. Charvet. 2013. A Meta-analysis of Supply Chain Integration and Firm Performance. *Journal of Supply Chain Management* 49(2): 34–57.

8. Mackelprang, A.W., et al. 2014. The Relationship Between Strategic Supply Chain Integration and Performance: A Meta-analytic Evaluation and Implications for Supply Chain Management Research. *Journal of Business Logistics* 35(1): 71–96.

9. Frankel, R., and D.A. Mollenkopf. 2015. Cross-functional Integration Revisited: Exploring the Conceptual Elephant. *Journal of Business Logistics* 36(1): 18–24.

10. Yan, T., and K. Dooley. 2014. Buyer–Supplier Collaboration Quality in New Product Development Projects. *Journal of Supply Chain Management* 50(2): 59–83.

11. Oliva, R., and N. Watson. 2011. Cross-functional Alignment in Supply Chain Planning: A Case Study of Sales and Operations Planning. *Journal of Operations Management* 29(5): 434–448.

12. Stank, T.P., T.J. Goldsby, and S.K. Vickery. 1999. Effect of Service Supplier Performance on Satisfaction and Loyalty of Store Managers in the Fast Food Industry. *Journal of Operations Management* 17(4): 429–447.

13. Germain, R., and K. NS Iyer. 2006. The Interaction of Internal and Downstream Integration and Its Association with Performance. *Journal of Business Logistics* 27(2): 29–52.

14. Oliva and Watson, 435.

15. Mollenkopf, D., A. Gibson, and L. Ozanne. 2000. The Integration of Marketing and Logistics Functions: An Empirical Examination of New Zealand Firms. *Journal of Business Logistics* 21(2): 89.

16. Kahn, K.B. 1996. Interdepartmental Integration: A Definition with Implications for Product Development Performance. *Journal of Product Innovation Management* 13(2): 137–151.

17. Kahn, K.B., and J.T. Mentzer. 1996. Logistics and Interdepartmental Integration. *International Journal of Physical Distribution & Logistics Management* 26(8): 6–14.

18. Kahn, Interdepartmental Integration, 137–151.

19. Ibid., 138.

20. Kahn, K.B., and J.T. Mentzer. 1998. Marketing's Integration with Other Departments. *Journal of Business Research* 42(1): 55

21. See Mello, J.E. 2013. Collaborative Forecasting: Beyond S&OP." *Foresight: The International Journal of Applied Forecasting*, 48: 6; J.E. Mello and R.A. Stahl. 2011. How S&OP Changes Corporate Culture: Results from Interviews with Seven Companies. *Foresight: The International Journal of Applied Forecasting* 20: 37–42; J.E. Mello. 2010. Corporate Culture and S&OP: Why Culture Counts. *Foresight: The International Journal of Applied Forecasting* 16(1): 4.

2

Bridging the Integration Gap

By David W. Schumann, Wendy L. Tate, and William A. Powell[1]

Within many companies, generating sales and executing operations to meet those sales requirements are largely disconnected; sales is accountable for generating revenue, whereas operations is responsible for cost management. Because of this disconnection, there are conflicting priorities, conflicting metrics, and many forgone business opportunities. For example, using promotions to generate sales without planning for those additional needs might create a bullwhip effect across the supply chain, causing inefficiencies, expediting, higher material prices, decreased quality, and potential labor issues within a manufacturing plant because of excess hours needed to meet the demand. Whereas communicating the potential demand issues with those on the supply chain side could eliminate (or at least minimize) the detrimental impacts. Bringing the supply and demand side of a business together can represent significant opportunity for efficiency, effectiveness, and value creation.

The divide between the demand and supply is a serious problem for organizations today. Although we have made great strides in technology, data access, and management capabilities, we still do not know how to effectively bridge the great divide. The real issue is the conflicting objectives between finance, marketing, and operations. The overarching goal is to maximize value to both customers and to the organization itself. The business press is replete with examples of misaligned objectives in typical day-to-day decision-making processes where the ability to communicate expectations and hear what others are saying is crucial for building the bridge. Consider, for example, two functions—marketing and logistics—discussing accountability for and need for inventory. Marketing is rewarded for being able to deliver products to its customers on time, meaning that available inventory would help them achieve their goals. Whereas logistics is rewarded for reducing inventory costs and increasing inventory turnover; not having excess inventory helps them achieve their goals. In this conflict, companies end up selling what is left over below the market price or losing sales because of not having enough inventory to meet demand. Is there a way to better manage this disparity?

Consider also a cross-functional team that is negotiating a new contract with a new supplier. Each individual on the team has specific objectives that often do not align. For example, if finance takes a hardball approach on negotiating price and other terms, this could damage purchasing's ability to develop a more strategic relationship with the supplier that would offer other benefits to the firm. In this same negotiation, engineering team members may prefer to "overdesign" with very tight specifications and potentially extremely high (and unnecessary) levels of quality (for example, Six-Sigma quality) when 95 percent quality with on-time delivery might work just as well. There are always trade-offs; higher quality equals higher price! Now the distinct needs of purchasing and finance are not being met. How can all the needs of an interfunctional team be understood and communicated in a way that achieves a win-win-win situation; customer, supplier, and your own organization?

Problems with labor relations can also be strained by this inability to bridge the supply- and demand-side gap. Imagine a situation where an analysis is being made to outsource some production requirements (find a supplier). This is a very involved decision that requires multiple functions (with conflicting objectives). Word gets down to the laborers that "they are going to lose their jobs because these items are being outsourced to an offshore location." The laborers can act in many ways: slow down production, go on strike, and/or ask for increased salaries (depending on the status of the labor contract). All these activities cause disruption in the supply chain and result in customers not receiving products on time. Inventory levels will go up as productivity goes down. Strained relations between management and the union workers could be avoided by better integrating demand and supply through the marketing, operations, and purchasing communications. But again, how can we effectively integrate these conflicting objectives to achieve the best overall outcome for customers, suppliers, and the firm?

The goal of integration is to achieve the right outcomes for the right process. This means that a clear and relevant mission or value focus is needed. What does the company want to communicate about itself, and how should it be going about communicating this focus? One of the biggest issues is how to share the relevant knowledge across the organization and also how to gather the relevant information needed for planning that leads to a decision. Aligning objectives that relate to the overall performance of the organization provides a foundation for employees to make decisions that maximize overall value. This also requires collaboration with others that may be "different" than in the past or may have different ways of communicating. This chapter focuses on this issue: integration is important, bridging the gap is crucial for many reasons, but *how* can we communicate and collaborate within and across an organization in a way that helps align and meet differing objectives and incentives to achieve the best overall value?

This chapter also seeks to describe and address the difficulty behind integration. *Interfunctional bias*, defined and discussed later, is believed to be a major cause of integration difficulty. The notion of a superordinate identity and accompanying conditions is reviewed as a means to decrease or omit interfunctional bias. However, the bulk of the chapter contains a multistep *interest-based collaborative problem solving* (IBCPS) process. This six-step process outlines a means for collaborating and communicating constructively and productively to overcome issues and problems assigned to cross-functional teams.

The Difficulty in Integrating

Demand and supply integration involves the cooperation of two or more functional workgroups. These groups are often distinct from one another in their training, responsibilities, and therefore, in their motives. For example, the sales force is typically incentivized to build customer relationships and ultimately make sales for the company. They are rarely motivated to sell customer orders that are not disruptive to the supply chain, nor are they motivated to participate in sales and operations planning. Simply, as noted earlier, their objectives are different from those of their supply chain colleagues.

Group behavior further complicates the challenge of cooperation and integration between two functional units. Groups refer to people finding common interests, goals, or similarity with other people, and identifying with these others who are like them. "In-group" members, or people who feel like they are a part of a group together, view out-group members, or people who are not a part of their group, with negative bias. As Henri Tajfel's pioneering research reflects, often without awareness, in-groups tend to hold positive attitudes and behave with favoritism toward their own group members. In contrast, the in-group tends to hold negative attitudes and behave unfavorably toward those who are not in their in-group, the "out-group." This dynamic has been found repeatedly in intergroup relationships.

Corporations often unintentionally promote such behavior through the formation of group structures and allocation of resources to groups based on the groups' functional responsibilities. For example, the supply chain employee who is forced to lose efficiency because of a customer order, or whose schedule is complicated by the request of a salesperson, may do more than simply comply. The employee's response may include forming negative attitudes toward a single salesperson or toward the sales force in general. These attitudes may be shared with others in the supply chain in-group. The response may include watercooler conversations with other supply chain employees about salespeople or other behaviors, such as a resistance to future requests, complaints to management, or even the sabotage of future disruptive orders. Clearly, a potential disrupter of demand and supply integration is the presence of biased attitudes and dysfunctional behaviors between distinct groups. Within corporate organizations, this is known as interfunctional bias.

Interfunctional Bias

Interfunctional bias operates in much the same way as racism and sexism. Racism can be broken down into a belief component known as stereotypes, an attitude component known as prejudice, and a behavioral component known as discrimination. Within corporations, different functional groups tend to hold general beliefs about other groups (that is, stereotypes), whether these beliefs are justified or not. For example, upon meeting a salesperson, it is likely that nonsalespeople (and perhaps some salespeople) will immediately recall beliefs and attitudes they hold toward salespeople in general. Regardless of the nature of the individual, they will be categorized and judged as perhaps aggressive, shallow, or driven only by personal wealth. Whatever the stereotype, salespeople and other functions carry with them the stigmas of their profession.

Interfunctional bias goes beyond the typical watercooler conversation about how one functional group is unfairly advantaged or uncooperative. Interfunctional bias suggests that negative attitudes toward another function can elicit uncooperative behaviors as well. Just as prejudice may spawn discrimination when race-based differences are the focus, prejudice toward another function can influence a break-down in demand and supply integration. Understanding the causes of interfunctional bias can inform managerial decisions about the potential remedies of this barrier to demand and supply integration.

The attempted integration of the demand and supply functions of the organization produces group-based beliefs, attitudes, and behaviors that contribute to the numerous challenges experienced within integration. Outgroup derogation can lead to less information sharing, inaccurate estimates and forecasting, unwillingness to consider another's perspective, or even the deliberate undermining of the other function's operations. Such behaviors may seem justified and even socially acceptable as biased group behavior goes unchecked or unnoticed. Just as race, gender, and age continue to be barriers to societal group integration, the functional distinctions of marketer, salesperson, operations or supply chain professional, purchaser, or logistician continue to be barriers to demand and supply integration.

General Methods for Overcoming Conflict Within Integration

Strategies for reducing intergroup bias and public policies that promote and attempt social integration are each worth understanding for how they might also apply in the corporate context. To begin, consider the usefulness of superordinate identity. Superordinate identity can be thought of as group membership that supersedes functional group memberships and that is inclusive of the entire corporate organization. When employees buy in to the notion that the entire company is one cohesive group that stands together and works together, boundaries and biases created by functional groups can be eliminated. Establishing and maintaining a superordinate identity is very challenging, especially for larger companies. The following four points highlight some of the challenges. These four points are integration-encouraging strategies adopted from research on the conditions under which disparate groups are more likely to integrate.

First, the literature on racial integration suggests that one important condition for successful integration is an equal status between the groups. Preliminary research on interfunctional bias suggests that equal status is highly correlated with the strength of stereotypes held by one functional group about another. The more equally members of different functional groups believe they are treated, rewarded, and *not* given priority, the less they will tend to hold biased beliefs about the out-groups. In business, inequality may be evident in resource allocation to one group over another, priority in decision making and negotiations, or even the dominant culture of the organization treating one function as more important than another. Such inequalities erode efforts toward demand and supply integration and undermine a sense of superordinate identity.

The second point to address is the outspoken support given to cross-functional integration by the leaders in the organization. Managers and supervisors can provide encouragement (and perhaps incentives) to work cooperatively with other functions. Supportive norms can then permeate throughout groups so that a normal mode of operation becomes one of working together. Demand and supply integration has not historically been a practice in business management, and so establishing and developing supportive norms in a culture where they have not previously been practiced may prove to be very difficult. Supportive norms are typically present in superordinate identity environments, given that in-groups support their members, favoring them over out-group members. Encouraging supportive norms between demand and supply functions seems important and practical, changing norms between groups can be a struggle.

Third, cooperative interdependence is especially applicable to interfunctional business relationships. A value chain has many links in the process of producing customer-delivered goods from raw materials. Each link in the chain relies on the preceding link for the value that it adds to the offering. This interdependence between functional units can be a powerful force in breaking down group distinctions. Not unlike an organizationwide superordinate identity, functional groups can experience camaraderie or cohesion through dependence that has historically been cooperative between the groups. In ongoing research on interfunctional bias, cooperative interdependence has been negatively correlated with interfunctional prejudice. More perceived interdependence between two groups results in fewer negative attitudes held by one group about another.

The fourth point of interest is the presence of interpersonal interaction between the groups. Although personal meetings between members from different groups might potentially reinforce stereotypes, the opposite is true. As members from distinct groups engage in meaningful interaction, the negative beliefs they may foster toward the other party are diminished. Within the corporate organization, having geographic distance or other physical barriers to interpersonal interaction can dehumanize members of the out-group and make holding biases about the out-group easier and more prevalent. Even within functional groups, cohesion can be fostered through face-to-face contact. Attempts at organizationwide group formation through superordinate identity would benefit from interpersonal interaction among all the employees, although this is often not physically feasible.

Clearly, these four points are not entirely independent from one another and likely interact with one another. Although it is possible to have cooperative interdependence despite a feeling of inequity between the groups, or to have supportive norms without interpersonal interaction, the existence of one likely influences the existence of another. In a company where all four conditions exist, a superordinate identity is more likely to be prevalent, and demand and supply integration is more likely to be unencumbered by interfunctional bias. If only one or two of these conditions are favorable between the functional workgroups, fostering the other conditions is advisable. At the very least, the higher degree to which disparate groups meet these four conditions that reduce interfunctional bias, the fewer socially derived barriers to demand and supply integration will exist.

Interest-Based Problem Solving and Collaborative Communication

Although the conditions that reinforce a superordinate identity must be considered in overcoming the difficulty of integration, what is arguably most important is how a cross-functional team communicates and solves problems; that is, how a team works together constructively and productively. Without a process for enhancing communication and consideration of disparate interests, efforts at working in cross-functional teams are doomed to failure.

Interest-based collaborative problem solving (IBCPS) is a process through which solution(s) are derived from a thorough dialogue about an existing problem or desired goal, the underlying interests of individuals, and a jointly constructed solution. Its focus is on understanding each person's or function's interests instead of focusing on agendas and hard/fast positions. It includes a set of useful steps for operating and communicating (Table 2-1). It invites respect, transparency, and open sharing of information. It constructively exposes functional cultures and processes. Within the IBCPS process, individuals share their motivations for reaching a productive solution, aspects of their own disciplinary identities, and constraints that exist within their respective environments. Working together, a diverse team determines a solution that aligns with objectives, processes, incentives, and satisfies individual needs as much as possible. The result is a formal proposal for change to provide decision makers and an assessment plan that examines the level of success of the proposed solution. The IBCPS includes six process steps. Each is described next in some detail.

Table 2-1 Steps of IBCPS

Step I: Establish Ground Rules/Conduct Initial Sharing/Determine Logistics

Step II: Develop and Reinforce a Constructive Communication Process

Step III: Define the Problem or Situation to Be Overcome/the Outcome to Be Obtained

Step IV: Identify Individual and Functional Interests

Step V: Consider Potential Actions and Select/Justify an Agreed-Upon Action Plan

Step VI: Provide a Cohesive Action Plan to Decision Makers and Other Stakeholders

Step I: Establish Ground Rules/Conduct Initial Sharing/ Determine Logistics

It is important to begin with a common understanding of how to operate within the team and a process for moving forward. This includes four substeps: (1) establishing ground rules, (2) understanding the nature of contributing functions (increasing awareness of each other's work), (3) determining logistics, and (4) determining the decision process.

After introductions (in settings where people are meeting each other for the first time), establishing ground rules for how the team will engage with each other is the initial step in putting the team on a path toward constructive interaction. A set of ground rules typically found in IBCPS is provided in Table 2-2. However, it is important that the team ultimately derive an agreeable set of rules in which to operate.

Table 2-2 Ground Rules for IBCPS

1. Treat everyone with legitimacy and respect.
2. Engage in active listening to others; be willing to participate actively.
3. Take responsibility for yourself, others, and the process.
4. Be honest and promote transparency.
5. Promote creativity.
6. Engage in demilitarized discussion.
7. Reinforce consensus decision making.
8. Assert your freedom to disagree when appropriate.
9. Commit to the (whole) process.

Understanding the nature of each other's work follows the setting of ground rules. This is a straightforward exercise in which participants describe to the group what they do, what their responsibilities are, and how they operate to get their jobs done. This sharing of roles and processes provides a common understanding of each other's potential contribution to an integrative solution.

The group also needs to agree upon logistical details in working as a team. This may include dates and times of meetings, location of meetings, initial materials needed, how long meetings should last, how notes will be recorded, technical resources needed, and so on. The team may also need to identify a chairperson or facilitator, a recorder, a person responsible for technology (including online distance communication), and the like.

Finally, the team needs to decide on a decision rule. A collaborative process works best when the group seeks consensus decisions. There are numerous consensus decision-making models available online. Consensus decisions take into consideration each group member's interests in solving the problem.

Step II: Develop and Reinforce a Constructive Communication Process

To truly integrate across functions, constructive communication is paramount. There are multiple models of enhanced communications. Here, borrowing from the numerous scholars on this topic, we recommend training employing a specific set of "collaborative communication" techniques. These include (1) questioning, (2) listening, (3) thinking, and (4) focusing. Each of these will be described in turn.

Questioning

A good question is perhaps the most important and rarely used technique in conversation. Too often we desire to share our own point of view, to advocate for our own position. Sadly, this minimizes listening to others. In collaborative communication, a balance between inquiry and advocacy is stressed. Indeed, it is often recommended that rather than advocating immediately (for example, stating a position or opinion), an individual construct an open-ended question (a question not answerable by one word, such as "yes" or "no") targeted to a person with whom they might disagree. In structuring an open-ended question, it is important to ask the question as objectively as possible without advocating for your position within the question. An example of an inappropriate question that includes advocacy is the following:

> Why do you think your solution has greater merit than the one we already have on the table? Your solution does not address a number of concerns I have that were included in the previous solution. Remember when I proposed...

This automatically biases the response one might receive and may put the other person in an unnecessary and counterproductive defensive position. Open-ended questions typically start with "Why," "What," and "How." They could also contain a scenario, followed by a question, the scenario posing a possible challenge to the initial perspective of the speaker.

A second type of question that is most helpful is "asking back." When a person asks you a question, after you answer them, ask them "Why did you ask this question?" or "What led you to ask this question?" or "What are you thinking about?" This "asking back" gets at the motivation an individual has to pose the question in the first place. By asking back, a new content enters the discussion. By using these tools, a true dialogue begins to occur.

Listening

Much has been written about "active listening." At the heart of good listening is the ability to suspend one's own thoughts while listening. Too often while something is being shared, a listener begins to generate a response. For example, someone tells a story, and the receiver starts focusing on a similar story that they want to share. Or someone argues for a perspective, and the receiver is quickly considering how to counterargue. What is happening in these cases is that the receiver has stopped fully listening to the speaker. We call this "reloading." When everyone is doing this at the same time, bringing in their own advocated positions at the same time, this is referred to as "popcorning." This is obviously highly dysfunctional.

In listening to others, one is attempting to listen, not only for content, but also the underlying assumptions, beliefs, and emotions of the speaker while suspending one's own assumptions, beliefs, and emotions. This is not easy to do and takes practice. In addition, in a group setting, it is important to be able to remember several threads back. One method of practice is to try to recall the content of the past five people. This allows the listener to not only identify each person's contribution, but it also affords an opportunity to tie these contributions together to create greater meaning.

Thinking

While you are suspending your thoughts while listening, it is also important to acknowledge that this "reloading" may be occurring and hold these thoughts for later, perhaps as you generate an open-ended question. This thinking will likely contain your own beliefs and assumptions that are important to consider in comparison to the perceived beliefs and assumptions of the speaker. Your thinking should not take the place of active listening, but nor should it be ignored altogether. This thinking is important to continuing the dialogue. This balance between listening, thinking, and questioning requires that the pace of the conversation be slowed, that pausing to process what is said and what will be said next is valued by the group.

Two tools are important here. First, when thoughts begin to occur while a person is listening to another, the individual needs to intentionally reinforce the following type of self-message: "I need to save this thought in my memory, but now return to the individual speaking." Second, when a thought takes over from listening and the person feels he or she has lost track, the person should not hesitate to ask the speaker to repeat something that was said. For example, "I'm sorry, I got lost there for a second; would you please repeat what you just said?" Or "Would you please go over that again? Thank you." In a sense, asking a person to repeat or review what was said gives others in the group the opportunity to pause, hear what was said again, and combine what was said into their own thinking.

Focusing

It is important in a group dialogue to stop at key moments and ask the question, "What are we talking about?" This keeps the group focused and alleviates the potential for popcorning. By asking each person this question, you might hear different perceptions of the dialogue. This is healthy because it exposes different lenses through which group members are processing the conversation. Focusing does not reflect what an individual is saying, but rather what the group is saying. What is the focus of the group at a specific moment? This focus will likely change as the dialogue proceeds, but stopping to review the content of the dialogue can be most helpful.

Training in Collaborative Communication

Collaborative communications training for an interfunctional team should ideally be led by a communications expert, one well versed in facilitating dialogue with collaborative communication tools. Training of this nature should never be overlooked. Too often, individuals come together as a group, each believing he or she is a good listener. Time and time again this has proven to be an incorrect and ultimately damaging assumption. The investment in such training can prevent significant conflicts and unnecessary and, ultimately, more costly delays. Moreover, such training repeated across cross-functional groups begins to reinforce a culture of open and productive communication.

Sharing Functional Information (Data, Processes, Policies, and Incentives)

Along with intentionally employing collaboration communication tools, proactive sharing of information is vitally important. This information takes the form of data, processes, policies, and incentives. By sharing these elements, different units will better understand the motivation of each unit's individuals, as well as the manner in which the unit operates. All too often, a breakdown in integration efforts occurs because of misaligned incentives or incompatibility of processes or policies. These need to be examined carefully and transparently to overcome constraints that may derail or erode the collaborative process. This sharing of information should occur whenever it is appropriate throughout the process. However, a specific place where it is typically critical is once the problem is defined (Step III) but before interests are shared (Step IV). After a problem is defined, having a common access to all relevant information may lead to a change in perspective, thus potentially influencing an individual's interests and needs.

Step III: Define the Problem or Situation to Overcome/ the Outcome to Be Achieved

The third step of the IBCPS process begins with constructing a shared definition of the problem. In some cases, an integrated team is presented with an already defined problem. In other cases, a team is presented with a desired goal, the problem of which is to develop a plan to reach that goal. A problem may also be defined in terms of a needed change. Regardless, a shared definition along with a discussion of the elements of the problem is important before going further. The team should begin by writing and editing the problem statement on a white board until everyone is in agreement. This edition process should reflect the concurrent discussion of the elements of the problem. After defining the problem, as noted above, a team must identify and gather information that will aid in subsequent work of the team. What is critical to the success of integration is a transparent sharing of information. The team needs to decide how this information will be disseminated (for example, by email, housed in an online shared storage program). Within Step III, teams need to adequately examine potential internal and external constraints. These may arise from the sharing of information. Internal constraints exist within a firm or supply chain and may reflect conflicts within existing processes or misalignment of incentives. External constraints reflect environmental influences that might include political and societal concerns, government policies and regulations, or present or forecasted economic conditions.

Step IV: Identify Individual and Functional Interests

Step IV reflects the heart of the IBCPS process. Based on the identified problem, individuals need to share their own motives for solving the identified problem or reaching a desired goal. Interests reflect individual and unit motivation and may be a function of incentives and/or professional ambition. Moreover, there are overarching interests of the company or the supply chain to be considered (such as fiscal responsibility, consistency with a corporate or supply chain vision and values). One means of sharing interests is to do an "affinity exercise." Each individual writes down one or more interests on Post-it notes (one interest per note). Everyone on the team places these interests on the board and then jointly categorizes these interests. Then each category is addressed, and individuals can pose open-ended questions about others' stated interests. In this way, a meaningful dialogue employing collaborative communication tools takes place and understanding of each other's motives and needs becomes clear to the group. Ultimately, the interests along with the ability to overcome constraints become the criteria in which the potential success of varied solutions (action plans) is evaluated.

Step V: Consider Potential Actions and Select/Justify an Agreed-Upon Action Plan

After individual and unit interests are identified and examined along with constraints, the team explores potential actions to resolve the problem. In so doing, the team explores multiple potential solutions. In generating solutions, the group may select to work in subgroups of two or three people. Each group generates multiple solutions for discussion. The team then dialogues about each action as it addresses the interests and ability to overcome constraints. The final solution derived should reflect as many of the members' interests as possible. It needs to be understood that not all interests may be able to be accommodated, thus requiring some level of individual and team compromise. Beyond identifying a solution, the team must also determine how the action will be implemented. What steps must occur and in what order? What processes must be reconsidered and revamped? Does the incentive structure need to be reconsidered given the desired action? What are the specific functional responsibilities that need to be carried out? Who is responsible? Is there a shared responsibility, and how will that be implemented? Finally, how will the plan be assessed in order to determine its level of success? What metrics will be employed?

Step VI: Provide a Cohesive Action Plan to Decision Makers and Other Stakeholders

After an action plan is agreed upon by the cross-functional team, a proposal is typically provided to the decision makers. The proposal should include a statement of the problem, a brief description of the process used to develop a solution itself, a thorough description of the solution, including stages as appropriate, the resources needed, changes that will be required (processes, incentives), and responsibilities with recommended assignments. Furthermore, the proposal should contain recommendations for communicating the action to various stakeholders (employees, customers). Finally, the proposal should contain a method with metrics for how to assess the success of the action plan.

As you can see from the preceding discussion, not only does IBCPS presented a step-by-step process for cross-functional groups to solve problems, it also provides a way of communicating that reinforces respect and transparency across different functions and disciplines and helps to bridge the great divide!

Conclusion

It is clear that better alignment of objectives that relate to the overall performance of the organization provides a foundation to make decisions that maximize overall value. Integration is important, but being able to clearly communicate those objectives and collaborate with others that have differing objectives is challenging. This chapter focused on a technique that can help us better align these objectives by communicating and collaborating within and across an organization. The idea is to be able to deliver overall organizational value and help bridge the gap between the demand side and the supply side of the organization.

Many companies are beginning to move toward a more integrated business model. However, this change requires adaptation to strategy and organizational structure. Designing and implementing incentives that measure overall organizational performance instead of functional-level performance are key. Using the process of IBCPS can help with the collaboration necessary for an integrated business model. It is difficult to break the mold of being an "operations-focused" organization or a "customer-focused" organization. For example, looking back at the issue of inventory, we can see that if an organization is operations or cost focused, it is doing all it can to minimize inventory, and it is not appropriately understanding the needs of sales in terms of promotions, seasonal fluctuations, or poor forecasting. Understanding the needs of both the demand and the supply sides requires a level of trust and a foundation for communication of the underlying business issues. There is no "right" or "wrong"—just strong differences in focus that need integration efforts. A customer-focused company might introduce so much variety into the production mix that purchasing and operations can't achieve any economies of scale or efficiency. Dialogue regarding standardized platforms or postponement strategies will increase the overall business results.

A team that is responsible for negotiation needs to use the IBCPS process prior to entering into the negotiation. The best results are those that are win-win-win, where all parties have discussed their expectations and the group can put a common face forward to achieve the best overall business results. In the movie *A Beautiful Mind,* John Nash discovered the roots of game theory by choosing a win-win strategy. Don't have everyone act in their individual interests, but instead act in the interests of the group—or in his terms, discuss the options and "don't everyone go for the blonde!"

The make or buy scenario is challenging because there are many hierarchical levels influenced by the results and many functional areas involved in the outcome. Using the IBCPS process to realize expectations and determine a process for accurate and transparent information will be necessary to address these issues. Integrating the demand and the supply sides of the business is extremely challenging. However, taking the first step and making a key strategic decision for a change in focus will start you on the journey. Finding a way to communicate this change in focus and address the different expectations will come only with clear and honest dialogue. This chapter presented a process that will facilitate dialogue. Developing and reinforcing a constructive communication process requires a natural learning curve as does becoming skilled in questioning, listening, focusing, and thinking in order to have dialogue between members of the organization with varying expectations. This is the key to being able to integrate the supply and demand sides of the business and realize better *overall* organizational value.

Endnotes

1. David W. Schumann is Professor Emeritus of Marketing, and Wendy L. Tate is an Associate Professor of Supply Chain Management in the University of Tennessee's Haslam College of Business. William A. Powell is a Haslam College alumnus and an Assistant Professor of Marketing at Shippensburg State University.

3

Maximizing Organizational Value Creation Across the Great Divide

By Diane A. Mollenkopf, Theodore P. Stank, and Wendy L. Tate[1]

Lack of a managerial orientation that integrates demand creation and supply fulfillment activities within firms causes them to be trapped in a pattern of reacting to the marketplace, selling excess product below market rates or losing sales due to insufficient product supply. Lessons from Fortune 500 companies provide insights into the radical changes in managerial thinking needed to integrate demand- and supply-side activities to create optimal value for both customers and the selling firm.[2]

Modern competition increasingly causes firms to compete in different product or customer categories that require them to develop capabilities in both cost leadership and product or service differentiation. Consumer packaged goods firms, for example, are adjusting traditional go-to-market strategies to provide omnichannel distribution. These firms must now sell their products across an array of consumer formats, each requiring unique supply chain capabilities. Such change, however, is not limited to the often-tumultuous consumer industries; even traditionally stable industries such as pharmaceuticals are undergoing transformation. Pharmaceutical firms, long accustomed to competing on product differentiation supported by patents, increasingly find themselves struggling to compete against generic providers as patents expire. They must seek cost efficiencies and determine how to deliver prescriptions directly to retail stores, or even to consumer homes and to consumers in the burgeoning emerging world market. At the same time, they must continue to innovate in new product categories such as biologics.

The ability to compete in different strategic segments requires precise coordination of marketing and sales efforts with operational delivery to ensure that appropriate value is created for each transaction. Yet, for many organizations, demand generation activities have become disconnected from the operational activities required to fulfill that demand. The disconnect results in unbalanced and often conflicting strategic objectives that Peter Drucker referred to as the "Great Divide."[3] Drucker attributed the Great Divide between the demand and supply functions within firms to be the reason they are so often trapped in patterns of reacting to a volatile marketplace, selling excess product well below market rates, or losing sales because there is not enough inventory of popular product available to meet demand.

Firms that suffer from the Great Divide do so chiefly because they have not proactively and strategically identified the best way to serve *customers of choice*—those strategic customers or customer segments that a firm purposefully pursues—in a way that is both valuable to the customer *and* profitable to the seller. Importantly, value requirements may vary significantly across customers or segments of choice, demanding cost efficiency in some cases and product or service differentiation in others. Marketing and sales personnel usually understand these value differences and develop product and service offerings appropriately. Yet most operational structures and processes used to fulfill demand are often incapable of varying their value delivery; thus, organizations find themselves trying to deliver efficient value to customers that want high levels of service, or vice versa. Marshall Fisher identified the mismatch of supply chain capabilities to desired value in 1997, yet the problem remains for most organizations.[4]

To overcome the challenge of the Great Divide, a radical change in managerial orientation is needed to integrate strategies and structures across both demand- and supply-side activities. Demand and supply integration (DSI) provides a means of facilitating the integration of organizational strategy with structure, and of connecting the voice of the customer to that of supply chain operations within organizations.[5] Interviews conducted with executives on both the demand and supply side of organizations suggest that DSI is a journey that starts with organizational leaders infusing a new way of thinking throughout the entire organization (see "About the Research"). Specifically, DSI involves coordinating the activities and processes related to a firm's ability to create demand with the operational, supply-side activities required to fulfill demand. DSI emphasizes the importance of leveraging market information and business intelligence to gain contextualized knowledge that informs organizations' strategic and operational decisions and commitments, which in turn enhances an organization's performance.[6]

The DSI journey starts with firms identifying customers of choice, understanding the value required by those customers, and developing capabilities to deliver on that value in such a way as to create economic profit for themselves; that is, the value that is relevant to both the customer and firm. As such, DSI enables firms to maximize the relevant value derived from serving customers of choice. The ability to balance the focus of demand creation with that of supply fulfillment promises to optimize the profit stream across all customers and customer segments pursued. Leaders of twenty-first century firms must help their organizations move away from a static focus on either cost or differentiation strategies. This chapter provides important insights to inform and guide the journey.

The Demand and Supply Integration Journey

Organizations have a long history of attempting to better integrate demand creation and supply fulfillment. The business press is replete with examples of various integrative decision-making activities, including Sales & Operational Planning; Collaborative Forecasting, Planning, and Replenishment; Early Supplier Involvement; and Design for Manufacturing/Design for Logistics. Such activities, however, are generally focused at functional levels of the organization and often fail to achieve desired results because they do not truly embrace the level and scope of change in overall managerial philosophy to ensure success. Too often, lower levels of the organization are left to "go through the motions" of integration without the appropriate mindset, knowledge, motivation, incentives, or structure to execute integration. Sales & Operational Planning, for example, often devolves from strategic-level allocation initiatives into operational "order status" meetings when strategic-level executives abandon the process over time and delegate attendance to lower-level functional managers.[7] The concept of Demand Chain Management (DCM), which involves managing the integration between demand and supply processes, the structure between the integrated processes and customer segments, and the working relationships between marketing and supply chain management, is silent regarding the relationship between the functional disciplines of marketing and supply chain management with top-level management and the changes in mindset and organizational structures and metrics required to incentivize the change to true DSI.[8]

To succeed, DSI requires top management to focus on aligning strategy and decision making with the activities that optimize relevant value, such as balancing appropriate value creation for customers of choice in a way that creates economic value for the selling organization. Findings from our research suggest that DSI maturity requires far more than merely implementing a single initiative; it cannot be implemented from the bottom up or over a short time span. Rather, it requires a programmatic push across all functions and organizational levels, from strategic to operational. Five distinct elements of DSI are required to fully conceive of and execute the delivery of relevant value. These elements include a Relevant Value Focus, which reflects strategic decision making and organizational design; Integrated Knowledge Sharing, which provides information support for decision making; Strategic Resource Prioritization and Integrated Behavior, which reflect supporting structural elements of processes and people; and Capacity and Demand Balance, which reflects operational execution.

Firms that have progressed the furthest toward DSI have done so in specific stages, beginning with a strategic-level emphasis and moving from there into the systematic development of structures, metrics, processes, and operational capabilities that ultimately enable the delivery of unique product and service bundles that optimize the relevant value for each transaction. Implementation of each successive stage represents a maturation process, as defined and presented visually in Figure 3-1.

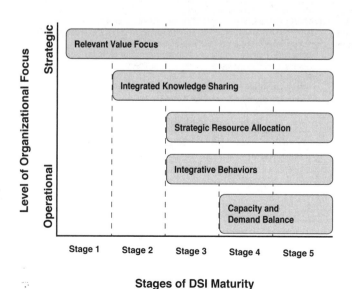

Stages of DSI Maturity

STAGE 1	**Relevant Value Focus**	The basis for a managerial orientation that shifts strategic focus to an emphasis on creating optimal value for both the focal organization and customers of choice. The first stage of an organization's path toward DSI; must be maintained throughout all stages to ensure success. *Requires:* • senior executive leadership must create the appropriate orientation towards value relevance • organizational structure that overcomes organizational silos • dashboard metrics to drive the business toward implementing DSI
STAGE 2	**Integrated Knowledge Sharing**	Enables implementation of relevant value focus by directing managerial attention to the knowledge networks needed to generate, disseminate, and interpret knowledge across a firm and customer/supplier networks to inform product/service value development. Each functional area (including customers & suppliers) should share knowledge puzzle to ensure the relevant decisions about how to create and fulfill the demand of customers of choice.[1]. *Requires:* • inter-functional collaboration within a firm • external collaboration with supply chain partners • adoption of the appropriate technology to facilitate collaboration
STAGE 3	**Strategic Resource Allocation**	Based on the knowledge created in stage 2, an enabling process addressing financial resource allocation for specific product/service bundles based on the needs of customers of choice, capacity constraints, and the capabilities of the firm that differentiate it from competitors. *Requires:* • tradeoff analysis and prioritization of resources toward initiatives that support customers of choice, and, importantly, away from non-prioritized customers and activities
STAGE 3	**Integrated Behaviors**	Another enabling process relating to the people who leverage the knowledge to integrate demand and supply. Employees must be educated, informed, empowered and rewarded in ways that enable them to create value relevance for customers of choice. *Requires:* • education and training • empowerment and accountability • evaluation and reward systems.
STAGE 4	**Capacity and Demand Balance**	The execution activities required to physically deliver the value promised by bridging demand and supply. Focus of operations changes from optimizing either cost or service to delivering relevant levels of product/service value to customers of choice at a profit given known capacity constraints. Emerges from successful implementation of stages 1-3, and represents the final stage of maturity towards DSI. *Requires:* • Effectively balancing demand from strategically important customers/segments with the operational capacity needed to deliver on those needs and requirements at a profit • Operational flexibility and fluid scheduling to meet demand variability and streamline processes • Coordination of capacity to move from an anticipatory to a more demand-responsive approach

Figure 3-1 The DSI Maturity Matrix[9]

Four Examples of the DSI Journey

Four Fortune 500 companies are profiled here to demonstrate the contrast between firms with only fundamental levels of DSI compared with those journeying toward DSI maturity. A diagnostic tool that measures each of the five elements of DSI (Relevant Value Focus, Integrated Knowledge Sharing, Strategic Resource Prioritization, Integrated Behavior, and Capacity and Demand Balance) was used to assess the maturity of the four companies, rating each on a scale of 1–10 (1 = poor; 10 = excellent) based on data compiled from interviews and summary reports of the individual organization (this diagnostic may be found at http://sloanreview.mit.edu/x/56403). Follow-up discussions with executives from each organization helped to validate the scores on each of the characteristics.

Two firms demonstrate low levels of DSI, with one biased toward a cost focus (Company A) and the other toward a differentiation focus (Company B). These firms are contrasted with two companies that enjoy a higher level of integration between demand- and supply-focused objectives that enable them to offer relevant value to different customers of choice (Company C and Company D). Although not one of the firms has yet perfected the integrated approach, each of the latter two firms has adopted a managerial orientation that transforms the way they do business. The latter two firms are beginning to see positive performance results from their efforts.

Each of the four highlighted companies was also compared to three of its major competitors on key financial metrics. The financial metrics selected for analysis, including cash-to-cash cycle, year-over-year change in cost and revenue, inventory turns, gross margin ROI, and return on assets, are often used in assessing both demand and supply performance.[10] The metrics are briefly described in Table 3-1 along with their role in assessing the performance of organizations in terms of the DSI framework.

Table 3-1 Financial Measures of Performance

Financial Metric	Description of Calculation	Business Usage	Determination of Performance
Cash-to-cash cycle	Accounts receivable days + average days in inventory – accounts receivable days	• This measure bridges the processes into and out of the firm. • Helps measure liquidity and organizational valuation.	• All industries are different; what is good in one may be bad in another. • Firms want to see the cash-to-cash cycle improving over time but not at the expense of the customer or supplier.
Year-over-year change in costs compared to YOY change in revenue	% change in costs compared to % change in revenue	• Provides understanding of what happens to costs as revenues change. • Measures whether a company is utilizing economies of scale.	• As revenues increase, costs likely increase, but ideally at a lower rate. • As revenue decreases, costs should also decrease, but at the same or greater percentage.
Inventory turns	Cost of goods sold ÷ inventory value (at cost?)	• Assesses how efficiently inventory is being utilized. • Provides insight into average days in inventory.	Inventory turns should continue to increase as demand and supply are better aligned.
Gross margin ROI	Gross margin ÷ by inventory turns	Measures a firm's ability to turn inventory into cash above the costs of the inventory.	A higher gross margin ROI contributes to higher valuation.
Return on assets	Net income ÷ total assets	Assesses firm efficiency of using assets to generate earnings.	A higher return on assets is a better indicator of success.

Company A: Automotive Assembly

Company A's extreme cost focus limits the firm's financial performance and overall progress. There is little attempt to identify desired customer segments or value creation targets, and the organizational structure is siloed with little cross-functional dialogue. Metrics are primarily cost focused, with little to no attention to service performance. Encouragingly, the operations side of the business shows signs of improving process management and collaboration, both internally and externally. However, there is a lack of leadership on how to incorporate these supply-side improvements with demand-side efforts to provide relevant value to customers of choice. Figure 3-2 reflects Company A's DSI maturity status, as well as the financial comparison of Company A to its three major competitors.

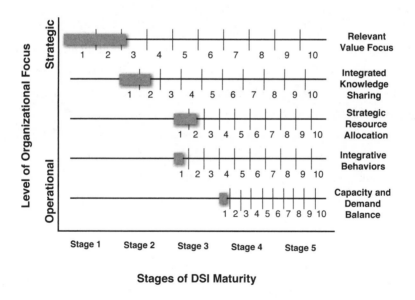

Figure 3-2 Company A, Too Cost Focused

Cash-to-Cash Conversion	• Poor; no improvement in last four years
Year-on-year Change in Costs/Revenues	• Extreme cost focus has enabled them to weather the economic storm; cost increases in line with recent revenue gains
Inventory Turns	• Average-to-poor, vis-à-vis competitors
Gross Margin ROI	• Average; has improved due to revenue growth, but cash-to-cash conversion rate suggests poor supply chain control
Return on Assets	• Below average

Financial evaluations conducted in comparison to the three leading competitors in the industry over the past four years. Actual numbers not included, ensuring anonymity.

Figure 3-2 Continued

With respect to knowledge sharing, internal collaboration is virtually nonexistent because the highly siloed nature of the organization allows little information or knowledge sharing across functional boundaries. The significant information technology (IT) within the organization is optimized within functional silos. A strong effort is placed on implementing additional technology, but without any underlying understanding of the processes the technology is to support. As one manager said, "We'll just have more islands of information," because the many existing systems are disjointed and nonintegrated, with little hope of being linked with new systems. External collaboration is "island-like" as well. One manager said, "Suppliers often have to develop their own forecasts because they can't get the information from us." The firm has little upstream visibility and struggles to manage the inbound flow of materials.

Financial resources are not allocated in a strategic manner, but based on functional cost minimization. For example, the company's low-cost sourcing policy achieves procurement objectives but increases costs related to quality control, delivery delays, and lifetime warranty support. A lack of overall strategy hinders the ability to prioritize resources or analyze trade-offs between options. On the personnel front, managers recognize the need for better training and development of employees, but cost constraints limit action. Lack of integrated knowledge systems means that people cannot get the data they need to do their jobs effectively. There is little accountability for data integrity or information accuracy, rendering reports inadequate. The forecasting process is equally in shambles, being based on financial goals, not market intelligence. Ultimately, there is no way to hold people accountable or to properly evaluate and reward behavior that would benefit the firm.

The extreme cost focus coupled with a lack of effective information systems inhibits this organization's ability to balance capacity and demand. The low-cost approach to dealing with suppliers compromises operational flexibility and capacity management. For example, trucking carrier management is based upon spot-market purchasing to ensure lowest cost. One manager admitted, "When carriers get paid better by other companies in today's economy, they quit our business and move others' freight." The localized information systems create blind spots within the company, negatively affecting inventory levels and customer service. There is little ability to coordinate capacity with the needs of the market, and even less operational flexibility to do so.

The performance metrics reflect the firm's lack of DSI maturity. Although the cost focus of this company allowed it to effectively weather the storm of the economic downturn, all other metrics are below the industry average. Costs are increasing at a rate greater than revenues, meaning that the company is not taking advantage of the economies of scale that become available as sales increase. Inventory turnover and return on assets are low, suggesting the company is not using its investment effectively. The GMROI is average; however, the cash-to-cash performance indicates that suppliers' payments are delayed and customer receipts are delayed, increasing the firm's financial exposure.

Company B: Health and Beauty Care Manufacturing

Company B's extreme marketing focus limits the firm's financial performance and overall progress. A lack of strategic direction provides limited guidelines for either demand- or supply-focused managers, reflecting a low level of relevant value focus. The firm is a market-driven, heavily branded company in which retail displays with a variety of product line offerings are critical for revenue generation and market share. There is no recognition at the senior level that the supply side of the business needs to be involved in the go-to-market planning and execution, resulting in significant inventory buildup. The strong cultural divide between the demand and supply sides of the company is particularly evident during new product launch events: more than 80 percent of new launches are late by at least 10 days, due mostly to the marketing and new product teams failing to inform the operations teams of schedules, quantities, and changes to plan. There is limited understanding of the profit impact of missing launch dates, including lost sales and expediting costs. Market share seems to be the only metric of importance, with little accountability for profitable sales. Somewhat promisingly, the company is in the initial stages of a sales and operational planning process designed to improve forecasting and better match marketplace demand with supply capabilities. Figure 3-3 reflects Company B's DSI maturity status, as well as the financial comparison of Company B to its three major competitors.

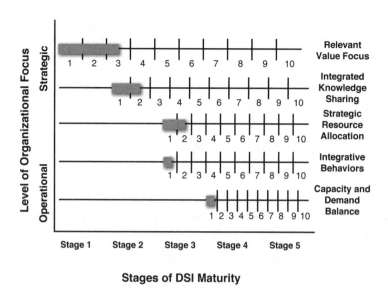

Cash-to-Cash Conversion	• Poor. Lots of variability in the past four years without improvement.			
Year-on-year Change in Costs/Revenues	• Good revenue increases over past four years, but no improvement in cost or inventory control			
Inventory Turns	• Poor; Extreme focus on revenue through SKU proliferation causes poor inventory performance			
Gross Margin ROI	• Average-to-Poor, relative to industry leader			
Return on Assets	• Good			

Financial evaluations conducted in comparison to the three leading competitors in the industry over the past four years. Actual numbers not included, ensuring anonymity.

Figure 3-3 Company B, Too Marketing Focused

Relevant information is not shared across functional boundaries in Company A. Information technology has been an underfunded area in the company, leading to inaccurate data and inconsistent reporting. Lack of good information means that the operations teams have little visibility into demand or inventory, making collaboration with upstream suppliers difficult. Supplier relationships tend to be more transactional than collaborative. The company is implementing an enterprise resource planning system that may help in improving accuracy and timeliness of data across the organization, which in turn will provide a platform for improved collaboration—both internally and externally.

The overall lack of focus on creating relevant value, coupled with poor information sharing capabilities and processes, means that strategic resource allocation is almost nonexistent at this company. Product proliferation suggests a poor prioritization of resources. Offering too many products can be counterproductive, confusing customers and increasing costs and complexity dramatically. One manager stated that "We introduce 1,000 new products every year, but with the resources we have, we can support only 25 percent of the new launches." A new regional manufacturing focus, designed to shorten time to market, is a move in the right direction regarding resource allocation. The localized factories should be tied closely to actual demand, thus enabling more discussions about available capacity and resource prioritization.

With respect to human resources, there is little emphasis on training and development of the operations and supply chain management personnel, often exacerbating the problems arising from poor information and a failure to prioritize resources. Little empowerment or accountability exists on either the marketing or operations sides of the business. Therefore, evaluation and reward systems do little to encourage strategic achievement of the organization's goals.

The current manufacturing and distribution network cannot adequately support the company's sales efforts; thus, sales are lost because of poor product availability. Complexities in the business (numerous brands and more than 30,000 SKUs) and in the supply chain (many suppliers in Asia with long lead times) make it difficult for the operational teams to meet the ever-changing needs of the market. This is especially critical for a fashion-based company in which flexibility and market responsiveness are important. Overall, operational flexibility and coordination of capacity across the network are at low levels, despite the new regional manufacturing focus mentioned previously.

With company B's focus on differentiation and its excessive SKU proliferation, performance in supply-and-demand metrics is impacted. The cash-to-cash conversion ratio has seen a lot of variability in the past four years. The company does well with its receivables; however, inventory turns and payables are both very slow. Revenue continues to increase, but costs are also increasing. Company B struggles to effectively manage inventory and subsequently suffers from very slow inventory turnover. The GMROI is consistently one of the lowest in the industry. Despite its struggles with inventory, Company B is doing a good job managing its longer-term assets that require additional investment. Many opportunities exist for improvement at Company B. The company has figured out how to manage the demand side but has done so at the expense of the supply side of the business.

Company C: Fast-Moving Consumer Goods Manufacturing

Company C is transforming to a DSI-driven organization. It is undertaking a radical philosophical change in business practices and is already seeing the effects of its new approach. A recent strategic change has led to a new focus on customers of choice; the company is developing a more strategic view in partnership with key customers about what the business should look like into the future. The company is now involved in joint business planning with its top customers. Managers focus on balancing supply chain capacity with the demand being created. This change, generated by the highest leadership, is a significant cultural change for the entire organization. A balanced scorecard approach embeds the new focus into the organizational culture. Common metrics across all functional areas support the new customer focus on relevant value. The company is in mid-transition but has a strong framework that is guiding decisions throughout the organization. Financial results are beginning to suggest the success of its new business approach. Figure 3-4 reflects Company C's DSI maturity status, as well as the financial comparison of Company C to its three major competitors.

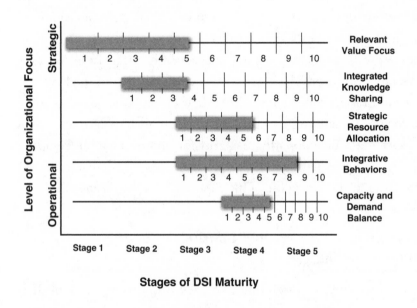

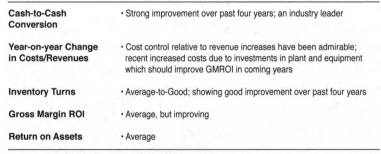

Cash-to-Cash Conversion	• Strong improvement over past four years; an industry leader
Year-on-year Change in Costs/Revenues	• Cost control relative to revenue increases have been admirable; recent increased costs due to investments in plant and equipment which should improve GMROI in coming years
Inventory Turns	• Average-to-Good; showing good improvement over past four years
Gross Margin ROI	• Average, but improving
Return on Assets	• Average

Financial evaluations conducted in comparison to the three leading competitors in the industry over the past four years. Actual numbers not included, ensuring anonymity.

Figure 3-4 Company C, Transitioning to a Balanced Focus

Joint business planning with key customers drives the market-based information necessary within the organization. Capabilities are nascent, but they are developing collaborative processes with key customers that will drive knowledge sharing for decision making. On the supply side, the organization is much less structured in working with suppliers in a collaborative manner. Previous efforts to collaborate had been poorly disguised cost-reduction approaches, but the new strategic approach is changing the way this company thinks about leveraging knowledge with its suppliers.

The new joint business planning processes are changing the way managers think about prioritizing and allocating resources, with marketing, sales and operations, and supply chain teams involved in planning and subsequent resource allocation. Resources are prioritized around key customers and jointly determined opportunities, balanced with capacity constraints. A more holistic approach to resource allocation allows the firm to make trade-off decisions more appropriately. Some initiatives that might not make sense from a purely functional perspective, such as the introduction of a new product with a particular grocery chain, may be recognized as part of a strategic growth initiative with that customer and thus be approved.

The balanced scorecard approach has become a regular part of the managers' business conversation in such a way that it is driving behavior at all levels of the organization. Integrated behaviors are driving higher levels of cross-functional knowledge sharing as well. As one manager stated, "We are living the balanced scorecard every day now." The training department has been instrumental in the balanced scorecard process, training and developing employees, and addressing retraining requirements to ensure that all employees are aligned with the overall direction of the organization.

With the new planning processes in place, a major emphasis is now being placed on execution. Better information about key customer profitability enables managers to make short-term decisions to flex their operational capacity while still making profitable sales. The joint business planning process focuses on a 12- to 24-month time horizon, enhancing day-to-day execution capabilities. In fact, the joint planning has reduced the need for reactive decision making that had dominated previous operations.

This company is well on its way to understanding what types of metrics best drive the business and how to better manage both the demand and the supply side of the organization. Company C has done an excellent job of managing its cash-to-cash conversion and has seen significant improvement over a four-year period. There have been significant investments in technology to improve the transparency and visibility of supply-chain-related information. This has increased costs in the short term but in the long term should radically improve inventory management. Today, Company C is above average on its performance metrics. However, the expectation is that following the DSI model with a focus on the relevant customers and the relative performance metrics will continue to improve the performance outcomes.

Company D: Consumer Electronics Manufacturing

Overall, Company D represents the most mature organization in terms of DSI implementation. It has made a significant cultural change in its organization and is well down the path toward achieving its goals. For Company D, proof of the transformation has been in the financial results achieved during the past several years. Figure 3-5 reflects Company D's DSI maturity status, as well as the financial comparison of Company D to its three major competitors.

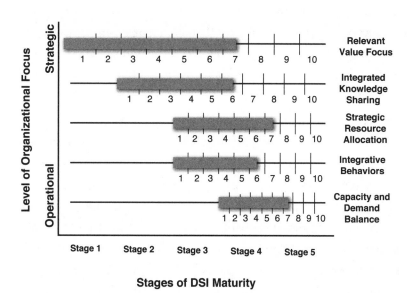

Cash-to-Cash Conversion	• Strong, and improving. Leading major competitors
Year-on-year Change in Costs/Revenues	• Good cost management. Costs rising slower than sales increases
Inventory Turns	• High turns relative to major competitors. Recent slowdown in turn rate reflective of a change in strategy, but still strong
Gross Margin ROI	• Strong and improving; an industry leader
Return on Assets	• Strong, leading all major competitors

Financial evaluations conducted in comparison to the three leading competitors in the industry over the past four years. Actual numbers not included, ensuring anonymity.

Figure 3-5 Company D, A More Balanced Focus

Historically, this firm chased market share at the exclusion of other goals and then had to accept significant amounts of obsolete inventory. A cultural change brought about by senior leaders has helped develop a more balanced approach. Emphasis is now on cross-functional alignment that spans customers of choice, new product development, and operational capacity decisions and activities. A balanced scorecard approach ensures that all managers are measured on organizational rather than functional goals. Revenue and cost are both important, but margin is the responsibility of both demand-focused and supply-focused managers.

Forecast accuracy is a major area of focus because it drives production planning. Past forecast errors led to trust issues across the demand and supply sides of the organization. Managers are working hard to develop a shared interpretation of knowledge across the organization based on one master sales plan and one production plan. External collaboration with suppliers is not as strong as senior managers would like. Previous relational approaches that reflected a constant focus on price concessions had compromised once strong relationships. This cost focus, exacerbated by the fast-paced nature of change in the industry, makes it challenging to build truly open, collaborative supplier relationships. The company is striving to make progress in this area and has a good supplier-relationship model in place with its 3PL providers that it can learn from to better manage knowledge across the supply chain network.

According to a senior manager, the organization has moved past a "make what we can sell" approach to one of "how best to serve each customer." Trade-offs are addressed by taking a holistic approach to understanding total costs relative to revenue potential. Both demand-side and supply-side managers are routinely involved in these conversations. Sales teams own the customer management processes but are also held accountable for customer profitability, which helps facilitate financial resource decisions. On the operations side, a cross-organizational focus has eliminated the pet-project approach of previous years.

Integrated behaviors represent the weakest link in the company's DSI transformation, although the need to invest in employee development is now receiving significant attention as managers seek to institutionalize the balanced approach to running the business. Managers believe that the incentive and reward systems are appropriate for driving integrated behaviors, as exemplified by holding sales personnel accountable for customer profitability. Evaluation metrics support decisions that enhance the firm's performance as opposed to functional-level performance. The firm is developing the training and development initiatives required to move the organization to a more mature footing.

The organization sees a nimbleness in its operational activities stemming from changes to its supply base, logistics operations, and network design that allows it to flex its operations and mobilize quickly as market opportunities emerge. Capitalizing on market opportunities, however, is based on an overall focus on customers of choice. Flexibility doesn't always mean "fast" to this company; rather, it means "appropriate" decisions for each customer so that day-to-day decisions are made with an eye toward capacity availability relative to key customer requirements. Daily execution has become much smoother in the balanced approach.

The financial performance metrics show that Company D is an industry leader. Recent strategy changes explain the slowdown in the movement of inventory as marketing channels have adjusted. However, all the metrics are strong and still improving.

Managerial Implications

The four cases highlight the managerial actions required at each stage of DSI implementation to achieve notable performance improvement. Examples are provided in Table 3-2 for Companies C and D, to demonstrate the path they are taking toward DSI. Their examples provide several key insights for managers, as summarized next.

Table 3-2 Relevant Value Focus

Recommended Managerial Actions	Fast-Moving Consumer Goods Manufacturing (C) Initiatives	Consumer Electronics Manufacturing (D) Initiatives
• Chart clear course and identify benefits. • Focus on value solutions, *not* products. • Hold leaders and functional champions accountable for change. • Understand external environment. • Identify customer values and internal strengths.	• Identified strategic growth accounts aligned with strategic objectives and internal strengths. • Focused new product innovation on speed to shelf/speed to market. • Refocused supply chain objectives on customer supply chain solutions. • Focus on people-process-technology.	• Focused on end-to-end transformation. • Focused on value solution delivery, *not* products. • Identified three customer segments: affordable selection, configurable product, and service solutions. • Identified value characteristics for each segment. • Dedicated resources to manage the change.

Integrated Knowledge Sharing

Does your organization focus managerial attention on the knowledge networks needed to generate, disseminate, and interpret knowledge across the organization and its customer and supplier networks to inform product and service value creation?

Recommended Managerial Actions	Fast-Moving Consumer Goods Manufacturing (C) Initiatives	Consumer Electronics Manufacturing (D) Initiatives
• Invest in integrated IT. • Seek collaboration with suppliers to improve information exchange. • Engage key customers to better understand value creation needs. • Improve business analytics capabilities.	• Enhanced retail customer collaboration to enable integrated customer solutions. • Enabled integrated IT across firm and supply chain through SAP. • Enhanced data-driven analytical capabilities to drive decisions for demand sensing and shaping.	• Applied demand shaping and customer collaboration to minimize cancelations and reorder impact. • Created vested supply partners to conform to new supply chain goals. • Established IT control tower to proactively manage continuity of supply issues—24-hour resolution.

Strategic Resource Allocation

Does your organization allocate resources to create products and services designed to optimize the needs of customers of choice, taking into account capacity constraints and differential capabilities?

Recommended Managerial Actions	Fast-Moving Consumer Goods Manufacturing (C) Initiatives	Consumer Electronics Manufacturing (D) Initiatives
• Create corporate executive champion. • Stratify product and service offerings. • Segment and prioritize customers by value desired and ROI to the firm. • Reduce product offering complexity.	• Focused net new resource allocation on developing new products and services for strategic account customers. • Established Joint Business Planning process run by executive committee. • Reduced overall product line offerings.	• Implemented Integrated Business Planning process owned by Corporate VP to align priorities and investments. • Created "one source of truth" demand process owned by Sale GM to create one number across sales, finance, and operations. • Simplified product and service offerings. • Expanded product life-cycle planning.

Integrated Behavior

Does your organization educate, inform, empower, and reward employees in ways that enable them to create optimal value for customers of choice while ensuring cash flows that generate economic profit?

Recommended Managerial Actions	Fast-Moving Consumer Goods Manufacturing (C) Initiatives	Consumer Electronics Manufacturing (D) Initiatives
• Engage and incentivize the entire organization to execute the plan. • Create standards to guide behaviors. • Train workforce to understand end goals of solution delivery and standard work routines.	• Worked with human resources to create key skills drivers to guide hiring and development. • Developed competencies for each job. • Invested in employee training and career development.	• Educated and incentivized employees to provide customers the desired flexibility. • Metrics focused on improved customer experience. • Created playbooks to manage key business events, supply chain disruptions, and product transitions.

Capacity and Demand Balance

Does your organization focus on the need to create appropriate levels of value for each transaction given known operational cost capacity constraints?

Recommended Managerial Actions	Fast-Moving Consumer Goods Manufacturing (C) Initiatives	Consumer Electronics Manufacturing (D) Initiatives
• Create end-to-end supply chain solutions. • Determine and engage supplier capabilities and capacity. • Reduce overall process complexity. • Implement lean to reduce waste and improve agility. • Design a segmented operation to support what customers value.	• Redesigned supply network to enhance system optimization. • Implemented lean run strategies at all operating facilities. • Partnerships formed with raw material suppliers to ensure FDA regulations/food contact regulations were met. • Partnerships with equipment suppliers to redesign support systems.	• Focused on solution delivery. • Executed lean; reduced waste. • Activated different supply chains for different customer/product pairings, including MTP, MTS, MTO, and CTO. • Leveraged multiple transportation choices, including ocean or air, direct or indirect, and customization or value-added services.

Performance Outcomes

Deliver relevant value; that is, the specific product/service bundle that creates economic profit for the customer and for the delivering organization.

Recommended Managerial Actions	Fast-Moving Consumer Goods Manufacturing (C) Initiatives	Consumer Electronics Manufacturing (D) Initiatives
• Enhance customer value delivery. • Reduce asset commitments. • Reduce operating expenses.	• Order cycle time from 10 to 4 hours. • First-hour production efficiency—95%. • Zero capital outlay. • Order fill rate improved. • Transportation stops reduce 50%. • Inventory from 50 to 25 days of supply.	• Offer next day or 2-day delivery in 14 countries through Build to Stock supply chain. • 43% of product line preconfigured to lower costs. • Work with suppliers to provide product and delivery requirements that meet specific needs. • Deliver full solutions for complex product configurations. • Increased revenue with existing customers.

Does your organization strive to optimize value creation for your customers of choice and cash flows that create economic profit (that is, relevant value)?

Run the Entire Business with a DSI Perspective

DSI is not a functional-level process, or even a process that exclusively focuses on integration between functional processes, but rather an organizationwide orientation that traverses company politics and functional domains to provide a basis on which to effectively and efficiently run an entire business enterprise. DSI requires a high level of organizational maturity to do it well. Many of the initiatives that firms have already employed (for example, S&OP) will be included as part of the DSI transformation, but they are not in themselves sufficient to lead an organization toward DSI maturity. Leadership is a key success factor in the DSI transformation. Leaders need to create the organizational climate for DSI and establish goals and metrics that align functions to organizational goals to enable the entire organization to create relevant value for customers of choice.

Guide DSI Decisions Through Information Visibility and Shared Knowledge

Organizations do not create value for themselves or their customers without closely collaborating across both internal functions and with their supply chain partners to share knowledge about customer value, supply chain capabilities, and constraints. This requires willingness to exchange information both within organizations and between partners to create a shared interpretation of knowledge that can be acted upon. Accordingly, arrangements made with partners become much more than simple buy/sell transactions, and include joint coordination and planning, and, most important, a willingness to share information and risk. This means that structures and processes need to be put in place to facilitate information flow and create visibility across the organization and its partners. Technology solutions may be a part of integrated knowledge sharing, but alone do not ensure integrated knowledge to support relevant value creation.

Prioritize Financial and Human Resource Deployment

Strategic direction is required to make smart resource choices to better manage supply and meet demand. Fundamentally, an understanding of which customers are the most important to the firm (that is, customers of choice) will help managers make the best decisions about how to invest their time and money in satisfying those customers. This means that trade-offs between meeting demand with available capacity will be needed so that priorities can be established. Without such prioritization, every decision within a firm becomes a territorial battle across the Great Divide. However, with prioritization, both demand and supply functions can make resource allocation decisions that impact which customers to serve and how best to serve them to create value for the customer and generate economic profit for the organization.

Similarly, empowered and accountable employees can make decisions that enhance DSI goals. The management of human resources is critical to achieving strategic objectives. Accountability tied to rewards and incentives will help ensure that the organization is pulling together toward creating value for customers of choice. Empowerment of employees to make decisions that are commensurate with their level and responsibility in the firm is also necessary. Empowerment must be coupled with informed knowledge so that employees can make decisions that best utilize firm resources while meeting requirements for customers of choice.

Align Operational Execution with Relevant Value Focus

Daily execution requires coordination between operations and demand-generating functions to deliver the required level of value to each customer on each transaction in such a way as to create economic profit. Flexibility and fluidity of process are required, fostered by internal alignment across operational activities as well as alignment with suppliers and customers to coordinate operations and together achieve a level of agility beyond that of competitors. Rules and work arrangements, as well as innovative performance measurement and reward systems, may be revised so that goals and objectives of partners are complementary and focused on joint achievement of the benefits each seeks from the relationship.

Conclusions

Significant barriers and challenges to DSI initiatives remain in areas such as functional integration, customer and supplier collaboration, performance metric alignment, and information connectivity that continue to prevent most organizations from achieving radical performance improvement. Progress in these areas has been, at best, incremental. Yet if an organization does things only incrementally better, it may expect only incremental improvement in key organizational metrics related to economic profit. Rather, organizations must continue to pursue radical innovation of thought and process to reach their full potential; radical improvement requires a radical change in managerial thinking. Pursuit of DSI promises one such pathway to radical improvement. The research reported here provides a road map for making the DSI journey, and also demonstrates that improved financial metrics may be expected along the journey.

About the Research

A qualitative case-based research design was employed to generate depth of understanding of the complex business phenomenon of integrating demand and supply activities within organizations. Because DSI represents a new way of thinking about managing for organizational success, our approach is appropriate for studying emergent practices.[11] In such cases, even small sample sizes can provide deep insights. Recognized research protocol was followed to ensure rigor in the data collection and analysis phases.[12]

In-depth interviews were conducted with managers and senior executives across a variety of functional areas in eight organizations to ascertain the different activities and processes that firms are developing to bridge the Great Divide. A set of open-ended questions was used to allow each participant the opportunity to share his or her experiences related to the management of supply-side and/or demand-side activities, as well as the facilitators and barriers surrounding the integration of various functional areas.

Interview transcripts were used to organize the extensive amount of data obtained during the interview process, compare findings, and develop trends and themes across the firms. Three members of the research team independently coded the transcripts to elicit themes from the participants' discussions. Codes were then compared and discussed; discrepancies were reviewed and analyzed until consensus was reached. Several phases of coding occurred, to elevate the *in vivo* codes representing the words of the participants to a more abstract level representing managerial concepts.[13] Ultimately, the codes coalesced around five themes that represent the elements of DSI maturity.

The results proved interesting in that the eight firms collectively reported many of what they considered to be the "right" initiatives needed to address the Great Divide between demand and supply management, yet there was conflicting evidence of successful integration across the eight firms. Secondary analysis, along with additional follow-up discussions with several of the firms, suggests the integration efforts must be structured appropriately and conducted within an appropriate managerial orientation. This discovery led to the Maturity Model and the Four Phases of the DSI journey as depicted in Figure 3-1.

Four firms in different stages of the DSI journey were then selected to showcase the DSI framework. The idea was to compare and contrast firms that have made a conscious effort to move toward DSI with two organizations that are focused on either demand side (differentiation) or supply side (cost focused). In order to compare the performance of these organizations, secondary financial data were employed. A number of metrics that are commonly used to assess both demand-side performance and supply-side performance were gathered. These metrics were then compared across the analyzed company and three other primary competitors. This was to understand the impact of DSI on financial performance. Table 3-1 has a list of the financial measures that were used and how they apply to DSI.

Endnotes

1. Diane A. Mollenkopf is the McCormick Associate Professor of Supply Chain Management, Theodore P. Stank is the Bruce Chair of Excellence in Business, and Wendy L. Tate is an Associate Professor of Supply Chain Management, all in the University of Tennessee's Haslam College of Business.

2. The research and results reported in this chapter were the subject of the following article: Tate, W., D. Mollenkopf, T.P. Stank, and A. Lago. 2015. Divided We Fall: Building Organizational Value Through Demand and Supply Integration, *Sloan Management Review*, Summer, 56(4).

3. Drucker, P.F. 2006. *Classic Drucker: The Wisdom of Peter Drucker from the Pages of the Harvard Business Review.* Boston: Harvard Business School Press.

4. Fisher, M.L. 1997. What Is the Right Supply Chain for Your Product? *Harvard Business Review* 65(2): 105–116.

5. Esper, T.L., A.E. Ellinger, T.P. Stank, D.J. Flint, and M. Moon. 2010. Demand and Supply Integration: A Conceptual Framework of Value Creation Through Knowledge Management. *Journal of the Academy of Marketing Science* 38 (1): 5–18.

6. Moon, M.A. 2013. *Demand and Supply Integration: The Key to World-Class Demand Forecasting,* Upper Saddle River, NJ: FT Press.

7. Ibid.

8. Jüttner, U., M. Christopher, and S. Baker. 2007. Demand Chain Management-Integrating Marketing and Supply Chain Management. *Industrial Marketing Management* 36(3): 377–392.

9. Crossan, M.M., H.W. Lane, and R.E. White. 1999. An Organizational Learning Framework: From Intuition to Institution. *Academy of Management Review* 522–537; Daft, R., and G. Huber. 1987. How Organizations Learn. *Research in the Sociology of Organizations* 5:1–36; Grant, R.M. 1996. Toward a Knowledge-Based Theory of the Firm. *Strategic Management Journal* 17:109–122; Huber, G.P. 1991. Organizational Learning: The Contributing Processes and the Literatures. *Organization Science* 88–115; Kohli, A.K., and B.J. Jaworski. 1990. Market Orientation: The Construct, Research Propositions, and Managerial Implications. *Journal of Marketing* 54(2):1–18; Narver, J.C., and S.F. Slater. 1990. The Effect of a Market Orientation on Business Profitability. *Journal of Marketing* 54(4):20–35; Nonaka, I. 1994. A Dynamic Theory of Organizational Knowledge

Creation. *Organization Science* 14–37; Sinkula, J.M. 1994. Market Information Processing and Organizational Learning.

10. Farris II, M.T., and P.D. Hutchison. 2002. Cash-to-Cash: The New Supply Chain Management Metric. *International Journal of Physical Distribution & Logistics Management* 32(4): 288–298. Randall, W.S. and T. Farris II. 2009. Supply Chain Financing: Using Cast-To-Cash Variables to Strengthen the Supply Chain. *International Journal of Physical Distribution & Logistics Management* 39(8).

11. Boyer, K.K. and M.L. Swink. 2008. Empirical Elephants—Why Multiple Methods Are Essential to Quality Research in Operations and Supply Chain Management. *Journal of Operations Management*, 26, 337–348; Flint, D.J., R.B. Woodruff, and S.F. Gardial. 2002. Exploring the Phenomenon of Customers' Desired Value Change in a Business-to-Business Context. *Journal of Marketing* 66 (4), 102–117.

12. Yin, R.K. 2009. *Case Study Research: Design and Methods*, 4th ed. Thousand Oaks, CA: Sage Publications, Inc.

13. See Charmaz, K.. 2006. *Constructing Grounded Theory: A Practical Guide Through Qualitative Analysis*. Thousand Oaks, CA: Sage Publications, Inc.

4

The Role of Information in Internal and External Integration

By Randy V. Bradley, Bogdan C. Bichescu, and Joon In[1]

With the increasing number of new technologies available in the marketplace, integration and standardization of these technologies have become increasingly more important to the overall effort to achieve better organizational outcomes and supply chain effectiveness. Along this line, several factors can inhibit and/or enhance an organization's ability to achieve its goals—information flow, information quality, and integration of IT infrastructure. Most major supply chain issues (in terms of execution) are related to a highly fragmented delivery system that lacks even rudimentary information management capabilities. This results in inadequate information flows and poorly designed processes characterized by unnecessary duplication of efforts, inordinate inbound and outbound lead times, and delays. Additionally, poor information quality is a major contributor to some widely known supply chain disruptions. For example, the supply chain glitches faced by Target Canada Inc., due to some barcode information that did not match the information stored in the information systems of Target's logistics contractor, resulted in errors in inventory levels at warehouses and delays in deliveries to stores. As a result, Target stores in Canada experienced understocked shelves, which in turn dissatisfied customers who visited stores. Although such issues may have occurred because of the mistakes of supply chain members (for example, buyers and vendors) and/or a glitch in their warehouse information system,[2] the more serious problem was that Target Canada Inc. could not identify and trace where the problem originated in the supply chain. In addition to exposing a lack of IT capability within organizations, such disruptions have created unprecedented demands on supply chain personnel to improve their IT acumen, even to the tune of designing, implementing, and managing large-scale IT integration projects.

Several factors enable supply chains' ability to consistently provide a high level of effectiveness—information flow, information quality, and integration of IT infrastructure. These factors, when done poorly or are at a low level, can inhibit supply chain effectiveness. Of the three factors mentioned, IT infrastructure integration is likely the most important, because the other factors are usually symptoms of IT integration problems. Furthermore, IT capability intentions are typically linked to IT infrastructure integration. In essence, integration doesn't just happen, and if it does, you probably don't want it. Thus, it is vital for an organization to adequately identify its goals and desires for IT capabilities to ensure that the investments in IT resources and integration are best suited to meet those goals.

The Importance of IT Infrastructure Integration to Supply Chain Effectiveness

IT infrastructure is a major catalyst for competitive advantage and sustained competitive advantage, especially in the context of supply chain management. IT infrastructure typically refers to the physical components, such as computer hardware and software (such as operating systems), network and telecommunications technologies, key data, core data-processing applications, and shared IT services, that reside or will reside in the organization. An integrated IT infrastructure is arguably one of the most important aspects of managing IT resources. Furthermore, an integrated IT infrastructure is the cornerstone upon which business activities in general, and supply chain activities in particular, are built. In fact, the growing strategic value of an integrated IT infrastructure is almost undeniable. One thing that makes Amazon's and Walmart's supply chains so nimble and effective is their tremendous investments in their IT infrastructure. This is not to say that the more you spend on IT, the better off you will be. Rather, it means the better you spend on IT (that is, investing in the right IT), for the sake of integrating the IT infrastructure, the greater the value derived from such investments. This also pertains to investing at the appropriate level, rather than arbitrarily throwing more money at IT. An appropriate level would be the level needed to attain the IT capabilities required to support chain activities.

Time is of the essence in virtually every industry, but regardless of the industry, it is absolutely critical in the supply chain. As such, IT infrastructure integration is particularly important for the supply chain, where access to information from anywhere at any time is crucial for effective and timely responses to changes in customer and supplier demands. Hence, a fundamental indicator of supply chain effectiveness is responsiveness.

One view of responsiveness is as a supply chain competency that enables organizations to react quickly to customers' and suppliers' needs and demands, without substantial cost increase. Responsive supply chains are able to adapt quickly to changing environmental conditions. Rapid adaptation can be critical when environmental conditions reflect tremendous ambiguity and uncertainty, as is often the case in high-velocity, dynamic environments with long, global supply chains. IT integration facilitates responsiveness by enabling faster interactivity and more seamless collaboration with both customers and suppliers. This interactivity and interorganizational collaboration involves the reciprocal exchange of information among all parties;

thus, errors and latency can be greatly reduced when this exchange occurs as quickly and smoothly as possible.

As the pace of change steadily increases in our global economy, business risk is compounded by rigid and nonintegrated IT infrastructures. IT infrastructure integration can enable organizations to easily, rapidly, and adequately respond to changes in their internal and external environments through the deployment and diffusion of IT resources and information assets. Further, IT infrastructure integration affords organizations the opportunity to implement and connect more complex IT resources (especially those that extend organizational boundaries), even as those resources and the environment evolve. As a result, supply chain partners can potentially have access to a more complete, consistent knowledgebase and technology toolkit with which to respond to needs of upstream and downstream members of a supply network.

It is well documented that an integrated environment is more than the sum of the individual components. Such an environment requires that standards be put in place to allow electronic communication between the various applications in the environment. When this is done, information can flow seamlessly from one application to another and, organizationally, from one function to another to allow for real-time decision making. Integrated environments have proven even more valuable when they are deeply rooted in the business processes of organizations.

The Role of Operating Models in Relation to Integration

Care should be taken to avoid overintegrating to the point that flexibility and nimbleness are replaced with rigidity and fragileness. This can be a trying "trial and error" process for some organizations, but the good news is that the guesswork can be removed. Researchers at MIT's Center for Information Systems Research[3] identified four operating models that organizations tend to follow (see Figure 4-1). An operating model is the necessary level of business process integration (BPI)—the extent to which business units and departments share data with each other—and business process standardization (BPS)— the extent to which business units and departments perform the same processes the same way—for delivering goods and services to customers. Hence the operating models are diversification (low BPI/low BPS), coordination (high BPI/low BPS), replication (low BPI/high BPS), and unification (high BPI/high BPS).

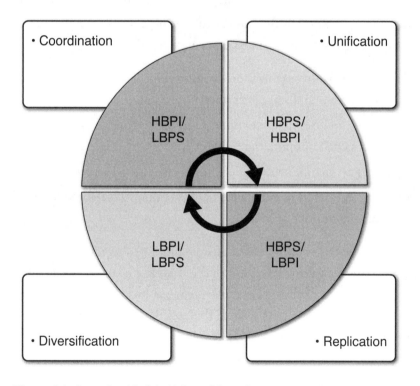

Figure 4-1 Operating Models (Adapted from Ross, et al. 2006)

Different companies have different levels of process integration. Integration enables end-to-end processing and a single face to the customer, but it forces a common understanding of data across diverse business units. Thus, companies need to make overt decisions about the importance of process integration. Key decision makers also must decide on the appropriate level of business process standardization. Process standardization creates efficiencies across business units but limits opportunities to customize services. Although various supply chain processes in a company may have different requirements for process integration and standardization, the key is committing to the way the organization, and thus its supply chain, will operate. Although most organizations have one operating model, it should be noted that some organizations follow multiple operating models simultaneously across different business units. It is tremendously valuable for organizations to go through the exercise of having key decision makers in multiple functions and business units answer the following questions:

- What is your organization's required level of business process integration and standardization?
- Which operating model is most representative of your organization's current operating model?
- Which operating model does your organization aspire to follow?
- Which operating model does it make more sense for your organization to follow?

Knowing the answers to the aforementioned questions and coming to terms with the gaps in responses from key decision makers can go a long way toward ensuring that organizations have the appropriate level of integration, as well as the ability to pivot transition to a different level, to ensure a high level of supply chain effectiveness continually.

Identifying the appropriate operating model and following through with it can help an organization avoid the pitfall of overinvesting in IT solutions. In fact, in our experiences, we find that leveraging the operation model to develop organizing logic for nurturing, developing, and creating IT capabilities that support effective execution of business processes is key to success. Along this line we have noticed an inverse relationship with respect to the number of supply chain IT applications implemented. In essence, organizations that build their foundation on the appropriate operating model tend to invest in fewer supply-chain-related technologies. Further, these types of organizations appear to have a more focused approach to building effective IT capabilities with both new and existing resources, instead of investing in technologies because of social contagion. Ultimately, investing in technologies for this reason typically means investing in something that may not be congruent with the organization's strategic orientation.

Conclusions

Supply chain and IT executives alike carry heavy decision-making burdens and must balance the need for more efficient and effective supply chain operations. With emerging technologies (for example, advanced robotics, embedded chips) and software hosting and delivery models available in the marketplace, the integration and standardization of the IT infrastructure will help to provide more access to relevant information. However, three salient questions need to be addressed pertaining to emerging technologies and cultural barriers, IT infrastructure roadblocks, and supply chain agility.

Issue #1: What are the most substantial technological and cultural barriers organizations are facing today?

With respect to this issue, we're not convinced that the most substantial barrier for organizations seeking to be more agile is technological in nature—at least not in the sense that requisite technologies or technological solutions don't exist. From our perspective, there is a cultural barrier that is a more formidable foe: "undercover heroism."

Undercover heroism is at work when an organization believes it's better than it truly is because it doesn't realize that it has heroes, behind the scenes, who are going to extraordinary lengths to not just do amazing things for the organization, but to accomplish relative minor success. Until organizations can come to terms with this sad reality, their continued attempts to achieve agile nirvana are futile. Shadow IT groups have not gone by the wayside; they're just much better at concealing their existence.

In the data processing era, the IT infrastructure mostly resembled a centralized computing infrastructure. This model eventually proved to be too tightly coupled and restrictive for the dawning of the "on demand" business landscape. As a result, IT infrastructures began transitioning to a more distributed architecture. Now, with more organizations expecting cloud infrastructure to be the key to their ascent to "Mt. Agile," we will ultimately realize that the more ideal path to an agile business is *"and"* not *"or."* In essence, we expect that organizations will eventually settle on what we refer to as centralized distribution. This approach has some infrastructure elements that are loosely coupled and others that are tightly coupled. Hence, it gives an organization an appropriate level of integration without a sentence of "life in rigidity."

Issue #2: What's the best way to identify roadblocks in your IT infrastructure that prevent your company from becoming more agile?

The best way to identify potential roadblocks in your IT infrastructure is to look downstream. By this, we mean consider the experiences of your internal and external stakeholders. How responsive do your stakeholders consider you to be when they make requests for information or services? How about when they make inquiries as to the status of various processes or activities? Do your stakeholders consistently receive accurate, timely, relevant, and consistent information, regardless of with whom they speak? Another way is to consider the efforts of personnel in your organization. How many disparate systems do they have to access to get to the "truth"? Do they spend a relatively enormous amount of time (on a consistent basis) aggregating data from those disparate systems and moving them into other systems for ease of analysis? These issues could very well be indicative of an IT infrastructure that is so loosely coupled that it inhibits the business from being agile.

Issue #3: Actions organizations can take that will lead them toward an agile, business-driven IT future

Stop overreliance on the cloud as a means of facilitating agility. The cloud can be an element in an organization's infrastructural scheme, but it should not be seen as a replacement for architecting a business that can sense and respond to changes (with relative ease and no substantial increases in costs) in its internal and external environment. Although business as a service (BaaS) sounds good in theory, the reality is that organizations often give up too much of who they are (for example, their core competencies) to effectively meet the needs of their stakeholders, especially those external to the organization. A key action is to strive for a level of IT infrastructure integration that doesn't result in rigidity. It's not about having every component of the IT infrastructure tightly coupled; it's having the appropriate elements coupled in a manner that facilitates the desired level of agility. Finally, recognize that every organizational strategy (no matter how good or bad) relies on a transactional framework that is undergirded by an IT infrastructure that enables agility or inhibits it.

As presented in this chapter, supply chain and IT executives would be well served to work together to clearly define the goals for IT capabilities in support of the supply chain function. When IT investment decisions are backed by clearly defined objectives, these investments are much more likely to result in more effective IT integration that will, in turn, improve customer-centric responsiveness and quality. It can be very easy to fall victim to the demands of entities and agencies external to the organization and to make hasty investment decisions based on the actions of competitors and the pressures caused by regulations, customer demands, and the like. However, in the longer term, organizations will be better served to consider these demands along with the goals of the organization in order to make the best investment decisions possible. Beginning with clearly defined goals for IT capabilities and then following that with investments in integrating their IT infrastructure will not only address the immediate pressures but will also provide organizations a means to achieve a high level of supply chain effectiveness.

Endnotes

1. Randy V. Bradley is an Assistant Professor of Information Systems and Supply Chain Management, Bogdan C. Bichescu is an Associate Professor of Management Science, and Joon In is a doctoral candidate in supply chain management, all at the University of Tennessee's Haslam College of Business.

2. Norton, S. 2014. Target's Canada Woes Highlight Supply Chain Shortfalls. http://mobile.blogs.wsj.com/cio/2014/05/22/targets-canada-woes-highlight-supply-chain-shortfalls/.

3. Ross, J.W., P. Weill, and D. Robertson. 2006. *Enterprise Architecture as Strategy: Creating a Foundation for Business Execution.* Boston, MA: Harvard Business School Press.

5

Bending the Chain: Deriving Value from Purchasing-Logistics Integration

By Theodore P. Stank, J. Paul Dittmann,
Chad W. Autry, Kenneth J. Petersen,
Michael J. Burnette, and Daniel A. Pellathy[1]

Over the past several decades, supply chain (SC) professionals have focused on performance issues that have emerged from a lack of commercial/business alignment with supply chain operations. Significant improvements have been made, and systemic processes like integrated business planning (IBP) and sales and operations planning (S&OP) have been developed to drive a fully integrated business. As business integration has continued to improve, the biggest supply chain opportunities have shifted.

Every year, the Haslam College of Business's Global Supply Chain Institute networks with hundreds of companies, requesting information on emerging supply chain issues. Our recent research shows that one of the greatest business integration opportunities is found within the traditional supply chain functions themselves. Specifically, we believe a major strategic integration opportunity exists between purchasing and logistics, and failing to capitalize on this opportunity is causing many firms to miss important opportunities to create value.

Based on our research, we believe it is probable that your firm is organized, measured, and incentivized in ways that essentially prevent you from deriving the full benefits of collaboration. In fact, it is highly likely that your company encourages behaviors that destroy value, both in the short term (by suboptimizing total system costs) and in the long term (by generating superficial gains from functional cost reductions while failing to leverage asset investments).

We have also uncovered strong evidence that organizations that align procurement and logistics decisions vertically with business unit strategy and horizontally between functions enjoy heightened levels of both functional and financial performance. In essence, these high-performing companies are able to *bend the chain* of plan, source, make, and deliver to enable alignment between purchasing and logistics. The result is that they serve customers better with lower operating expenses, cost of goods sold, and inventory.

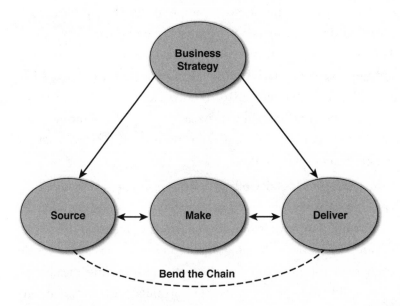

Figure 5-1 Strategic Alignment

Our research also sheds light on the structures, processes, and tactics top firms employ to enable this type of functional integration. Data from more than 180 supply chain leaders (firms ranging in size from over $20 billion to under $100 million) were collected and have allowed us to draw the following high-level conclusions:

- Procurement and logistics frequently are found in a broader supply chain or operations organization, but really exist as two separate and disconnected functions.

- Both procurement and logistics are well aligned independently to their business unit's strategy and activities but not nearly as well aligned to each other.

- Despite formal organizational links between purchasing and logistics, interaction between these functions is typically informal and unstructured.

Similarly, in our own experience we have found that when functional elements of the supply chain align with each other, improvements in firm financials and earnings per share invariably follow. Without integrated decision making, financial performance is at best suboptimized and at worst destructive to value. Firms must refocus organizational design, metrics, talent, and incentives to align activities across the value chain.

Finally, we conducted an analysis to determine whether the data provided any indication as to whether procurement and logistics integration (referred to as PLi in this chapter) was perceived as being an important lever of overall business success. The data clearly show that *integrated purchasing/logistics organizations deliver better business results* (that is, cost productivity, working capital productivity, and product availability).

Additionally, the many interviews we conducted with leading supply chain firms clearly suggest that companies with "best-in-class" supply chains consistently deliver the strongest business results. These best-in-class organizations tend to employ a set of four best practices:

- Fully integrated end-to-end supply chain organization integrated with common metrics.

- A talented supply chain organization that rewards people for in-depth mastery and end-to-end supply chain leadership.

- A purchasing and logistics network with an operating decision framework based on best overall total value of ownership (TVO; total cost of ownership plus level of customer value creation).

- Effective information systems and work processes that enable superior business results by providing multifunctional supply chain teams with the proper tools and information.

Finally, through our research and best-in-class interviews, we have been able to define a short list of actionable steps supply chain leaders can take today to make a difference.

1. *Get it on business leader scorecards.* Change the business reward system and culture from "suboptimal functional goals" to "total value creation for the enterprise."

2. *Champion TVO.* It is not enough to talk about use of total value of ownership with your direct reports. Personally lead the change in the supply chain.

3. *Make R&D your best friend.* Create a seamless technical community that is aligned on total business value creation between R&D and supply chain. New product supply chain design should be a seamless technical community deliverable.

4. *Set clear expectations on the use of multidiscipline teams in analysis and decision making.*

5. *Champion an end-to-end and integrated supply chain organization.* In the short term, align on a common direction if the purchasing and logistics teams have different leadership. Ensure that both organizations have a common supplier direction, scorecards, and rewards.

6. *Build supply chain talent that includes end-to-end supply chain mastery.*

7. *Partner with finance.* Work with finance leadership to align on how your multidiscipline teams quantify value for quality, customer service, environmental, sustainability, delivery, cost, and inventory.

The Surprising Challenge: Purchasing and Logistics Integration

Quality management icon W. Edwards Deming asserted more than 30 years ago in the first of his famous 14 points that a business enterprise needs constancy of purpose to succeed. Without this consistency of purpose, the business is not an organization but rather a collection of functions acting in disjointed and contradictory ways, impeding or even destroying value. Obvious improvements cannot be implemented, and ultimately business activities fail to create a chain that produces value for the company and its customers. Deming's solution to this fundamental problem was to focus on overt collaboration between functions. However, as our research shows, most companies still fail to follow through on his prescription.

Instead of adopting this advice, all too often organizations have focused on developing technical centers of functional expertise to drive scale and meet short-term financial and market expectations.

In the past five years, we have conducted more than 700 interviews with managers across all industries as part of the University of Tennessee's College of Business supply chain audit program. At the end of every interview, we always ask a "wish list" question: if you could change the world, what would you do to improve things in your company? By far the most common answer to that question is the desire for all the functions in the company to work together and become perfectly aligned toward a common purpose. People we interview pine for an environment where the functional silo walls have come down. They intuitively know that these disconnects are the real reason things are not improving faster.

In this chapter we discuss the results of a large-scale research initiative, along with real-life industry examples, which point to the fact that collaboration across functions and between enterprises is woefully missing from the value chain practice despite at least three decades of focus in the popular and academic press. More important, we show that when processes are integrated and silo walls are eliminated, the results can be very significant.

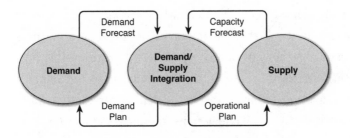

Addressed by: S&OP, SIOP, IBP

Figure 5-2 Integration Across the Supply Chain

As research by consulting firm Oliver Wight has already shown,[2] when companies integrate, the following changes result:

- Revenue goes up 10 to 16 percent.
- Fill rates go up 10 to 48 percent.
- Logistics costs go down 10 to 32 percent.
- Inventory goes down 15 to 46 percent.

Similarly, in our own experience we have found that when functional elements of the supply chain align with one another, improvements in firm financials and earnings per share invariably follow. Without integrated decision making, financial performance is at best suboptimized and at worst destructive to value. Firms must refocus organizational design, metrics, talent, and incentives to align activities across the value chain.

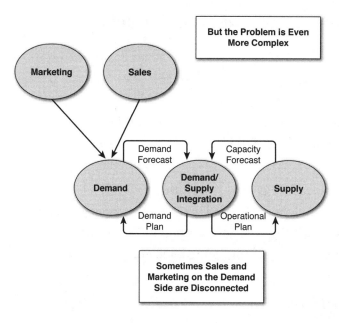

Figure 5-3 Demand-Side Disconnects

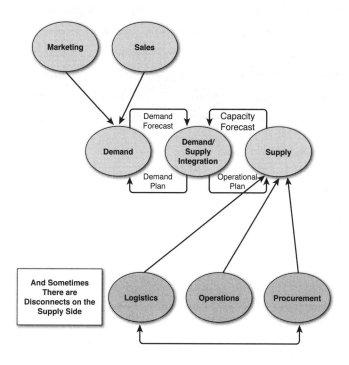

Figure 5-4 Supply-Side Disconnects

Supply and Demand Disconnects

For years, supply chain leaders have debated and discussed the disconnect between the supply and demand sides of business organizations. This lack of integration between sales and operations has spawned entire industries around ideas like S&OP; sales, inventory, and operations planning (SIOP); and integrated business planning (IBP).

But the disconnects go far beyond this macro level. For example, there is often a lack of integration within the demand side of the firm. Sales and marketing in manufacturing companies are not always aligned, nor are sales, marketing, and merchandising in retailers.

Supply-Side Disconnects

Similar disconnects occur on the supply side among logistics, operations, and procurement. Our research shows that one of the greatest opportunities for "lack of integration or dis-integration" occurs within areas traditionally thought of as supply chain functions. Specifically, we believe a major strategic integration opportunity exists between purchasing and logistics, and failing to capitalize on this opportunity is causing many firms to miss important opportunities to create value.

The Surprising Gap Between Purchasing and Logistics

Ideally, the supply chain functions of plan, source, make, and deliver are aligned and focused on serving the customer while simultaneously delivering world-class cost and working capital levels. The two functional areas of purchasing and logistics each have a major impact on these goals. Together, purchasing and logistics can represent up to 70 percent of total organizational costs and influence 80 percent of working capital through inventory and payables. Yet decisions made in these two areas are rarely made in concert with each other. In fact, purchasing often focuses decision making on optimizing metrics associated with purchase price and cost of goods sold, whereas logistics is focused on optimizing metrics associated with delivery and storage efficiency and effectiveness. Neither area tracks performance to higher-level financial value creation.

Example

A logistics executive for a large global consumer durable goods company hosted a "supply chain management advisory board." During dinner at a local restaurant, the executive leading this group noticed another group from their firm with a group of visitors in another private room in the restaurant. It turned out this other group consisted of the company's purchasing executives hosting their own "supply chain management advisory board." Neither group, to their collective surprise and chagrin, had any knowledge that the other group was meeting, nor did one know what the other was talking about.

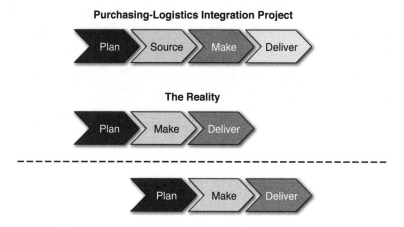

Figure 5-5 The Gap Between Purchasing and Logistics Models

What is the takeaway from this story? As both sides thought about it, they realized that it was symptomatic of a purchasing group making decisions about purchasing locations globally with no insight into costs of movement. At the same time, the logistics group was focused on how to reduce costs of global warehousing, inventory, and transportation with no insights into future locations of supply and manufacturing. The ideal plan-source-make-deliver model morphed into a new disconnected reality.

The Research: Linking Purchasing and Logistics Integration (PLi) to Improved Functional and Financial Performance

A survey was sent to purchasing and logistics managers from the University of Tennessee Global Supply Chain Institute and Forums mailing list, resulting in more than 180 responses from managers, ranging from CEOs and presidents to analysts. The respondent firms ranged in size from over $20 billion to under $100 million in revenue. The industries included the following:

- Aerospace/defense
- Apparel/textile
- Automotive
- Building materials
- Chemical, oil, and gas
- Commercial printing
- Components and systems
- Conglomerate
- Construction
- Consumer electronics
- Engineering
- Environmental services
- Facilities management services

- Financial institutions—banking
- Financial institutions—insurance
- Food, beverage, and nutrition
- Food service
- Government—national
- Government—local
- Health-care delivery services
- Heavy machinery
- High-tech network infrastructure
- Hotel/hospitality
- Household, personal care, and cosmetics
- Industrial equipment
- Media/entertainment
- Medical equipment
- Metals/glass processing
- Mining
- Office equipment
- Packaging
- Pharmaceuticals
- Plastics processing
- Professional/information services
- Pulp and paper
- Retail
- Telecommunications services
- Transportation services
- Utilities
- White goods

Respondents were first asked to identify whether they worked primarily in purchasing or in logistics. Purchasing was defined as including the following:

- Sourcing direct materials
- Procurement of maintenance, repair, and operating supplies
- Contracting services with outside suppliers
- Procurement of capital equipment/facilities
- Procurement of finished goods (completed items for resale)
- Supplier evaluation and selection
- Management of continuous supplier relations
- Supplier performance measurement
- Establishment of goods/services specifications
- Contract negotiations over materials supplies/services
- Global sourcing/sourcing strategy

Logistics was defined as including the following activities:

- Inbound/outbound transportation
- Owned fleet management
- Warehouse operations management
- Materials handling
- Packaging
- Order fulfillment
- Logistics information systems management
- Inventory management
- Management of third-party logistics services providers
- Customer service
- Reverse logistics flows
- Supply/demand planning

Next, the respondents were asked a series of questions related to their perspective on the nature and level of integration between their department and overall business strategy as well as between the purchasing and logistics functions. For example, if respondents indicated they were purchasing managers, they were asked about the purchasing group's alignment with business strategy and the group's relationship with the logistics group.

Major findings from the survey include the following:

- Purchasing and logistics frequently are found in a broader supply chain or operations organization but really exist as two separate and disconnected functions.

- Both purchasing and logistics are well aligned independently with their business unit's strategy and activities but not nearly as well aligned with each other.

- Despite formal organizational links between purchasing and logistics, interaction between the functions is typically informal and unstructured.

- Maintaining open lines of communication is the most widely supported method of interaction between the functions.

More detail on these findings is provided in the tabular breakdowns that follow.

Major Finding 1: Purchasing and logistics frequently are found in a broader supply chain or operations organization but really exist as two separate and disconnected functions. (Table 5-1)

While nearly 58 percent of respondents reported that purchasing and logistics were part of a common supply chain organization, more than 45 percent felt that they exist as separate functions. Fourteen percent still viewed purchasing and logistics as separate functions that are not part of the same supply chain organization, and 28 percent reported some other organizational structure.

Table 5-1 Procurement and Logistics Are Often Disconnected

Which of the following best describes the organizational structure for purchasing and logistics?	Percent Responding
Procurement and logistics are separate functions and are not part of a common supply chain organization	14.0%
Procurement and logistics are separate functions but are part of a common supply chain organization	45.5%
Procurement and logistics are part of the same function and are part of a common supply chain organization	12.2%
Other/Not Applicable	28.4%

Major Finding 2: Both purchasing and logistics are well aligned independently with their business unit's strategy and activities but not nearly as well aligned with each other. (Table 5-2)

The respondents provided a very strong indication that both purchasing and logistics functions are well aligned to business unit strategy and activities. That means both groups essentially agreed with the statements supporting the alignment of purchasing and logistics with business unit strategy (1 = strongly disagree and 5 = strongly agree).

Table 5-2 Extent to Which Purchasing and Logistics Are Aligned with Business Strategy

My Functional Area:	Purchasing	Logistics	Total Sample
Identifies opportunities to support the company's strategic direction	4.28	3.99	4.08
Understands the strategic priorities of the company's senior leadership	4.17	3.98	4.03
Adapts its strategy to the changing objectives of the company	4.21	3.89	3.99
Adapts its activities/processes to strategic changes	3.96	3.85	3.89
Maintains a common understanding with the company's senior leadership on its role in supporting strategy	3.92	3.70	3.77

My Functional Area:	Purchasing	Logistics	Total Sample
Educates the company's senior leadership on the importance of procurement/logistics activities	3.72	3.63	3.66
Assesses the strategic importance of emerging trends in procurement/ logistics for the company	3.60	3.51	3.54

Major Finding 3: Despite formal organizational links between purchasing and logistics, interaction between the functions is typically informal and unstructured. (Table 5-3)

Respondents were asked the level of engagement with the other function through a series of questions, where 1 = strongly disagree and 5 = strongly agree.

Of the different ways that purchasing and logistics might engage, informally working together, sharing ideas and information, and working together on a team scored the highest. More proactive approaches to collaboration, such as anticipating operational problems together and sharing resources, were by far the lowest. This supports the belief that purchasing and logistics, even when housed in the same supply chain organization, continue to operate in their own siloed worlds. Interestingly, purchasing managers perceived a much higher level of engagement.

Table 5-3 Level to Which Purchasing and Logistics Interact

My Function Engages the Other in the Following Ways:	Purchasing	Logistics	Total Sample
Informally working together	3.60	3.52	3.55
Sharing ideas and/or information	3.70	3.46	3.53
Working together as a team	3.77	3.42	3.53
Resolving operational problems together	3.75	3.38	3.49
Achieving goals collectively	3.58	3.30	3.39

My Function Engages the Other in the Following Ways:	Purchasing	Logistics	Total Sample
Developing a mutual understanding of responsibilities	3.64	3.29	3.39
Making joint decisions about ways to improve overall operations	**3.62**	**3.16**	3.30
Anticipating operational problems together	3.32	3.12	3.18
Sharing resources	3.30	2.98	3.07

Major Finding 4: Maintaining open lines of communication is the most widely used technique to foster integration. (Table 5-4)

When respondents were asked how purchasing and logistics interact, maintaining open lines of communication emerged as the most important technique. These open lines are informal and typically not systemic. Again, more proactive approaches, such as identifying potential sources of tension and establishing joint prioritization of projects, were ranked lowest (1 = strongly disagree and 5 = strongly agree).

Table 5-4 How to Foster Integration Between Purchasing and Logistics

Purchasing/Logistics Group Tends to Work with the Other in the Following Ways:	Purchasing	Logistics	Total Sample
Maintaining open lines of communication	3.94	3.52	3.65
Combining efforts on major initiatives	3.72	3.47	3.54
Developing clear lines of managerial responsibility for implementing plans	3.38	3.24	3.28
Achieving a general level of agreement on risks/trade-offs among projects	3.43	3.20	3.27
Coordinating project development efforts	3.53	3.16	3.27
Addressing potential sources of tension between procurement and logistics	3.21	3.14	3.16
Establishing a joint basis for prioritizing projects	3.28	2.98	3.07

We also asked respondents to indicate their functional area's performance relative to expectations, where 1 = well below expectations and 5 = well above expectations (Table 5-5). Not surprisingly, purchasing managers felt their performance relative to expectations was greatest for performance metrics over which they have the most control, such as performing to purchase price/cost objectives, supplier quality, payment terms with suppliers, and supplier responsiveness/flexibility. Performance metrics that require collaboration with logistics to achieve were all well below 3.0 on the 5-point scale.

Table 5-5 Purchasing's Performance Relative to Expectations

My Purchasing Group's Performance Compared to Expectations for Each of the Following:	
Performing to purchase price/cost objectives	3.28
Supplier quality	3.26
Payment terms with suppliers	3.17
Supplier responsiveness/flexibility	3.11
Supplier on-time delivery	2.87
Total cost of ownership	2.83
Supplier technology contribution	2.57
Inventory investment cost for purchased goods	2.40
Transportation and logistics costs	2.40

Similarly, logistics managers felt their functional performance exceeded expectations on metrics related to customer delivery. For example, customer service level establishment; network design/network location; full, damage-free, and on-time deliveries; and inbound/outbound transportation contracting and management are all metrics that fall under their control (Table 5-6). Performance metrics that require collaboration with other areas of the supply chain (for example, forecasting accuracy, total inventory turns, reverse logistics management, and time on back order) were among the lowest scores in the entire survey.

Table 5-6 Logistics's Performance Relative to Expectations

My Logistics Group's Performance Compared to Expectations for Each of the Following:	
Establishing customer service levels	3.59
Network design/network location	3.38
Full, damage-free, and on-time deliveries	3.19
Inbound/outbound transportation contracting	3.16
Inbound/outbound transportation management	3.07
Inventory planning	2.92
Logistics information systems design and implementation	2.86
Transportation costs	2.72
Total logistics costs	2.70
Time between order receipt and delivery	2.53
Warehousing costs	2.47
Logistics performance measurement	2.43
Line-item fill rate	2.34
Inventory costs	2.18
Order fulfillment management	2.03
Finished goods inventory	2.00
Forecasting accuracy	1.98
Total inventory turns	1.93
Reverse logistics management	1.89
Time on backorder	1.63

Finally, we conducted an analysis to determine whether the data provided any indication as to whether PLi was perceived as an important lever of overall business success (Table 5-7). Although this statistic is highly subjective, the table that follows provides indications that managers from firms in the top 25 percent of PLi in this survey believe their firms significantly outperform their competitors as compared with managers from firms with lower PLi scores (1 = well below competitors and 5 = well above competitors). In other words, *managers believe their company achieves a significant performance premium from aligning their purchasing and logistics functions.*

Table 5-7 Perceived Alignment of PLi to Business Success

My Firm's Performance in Comparison to My Competitors	Purchasing and Logistics Alignment		
	Firms in Top 25% of PLi Scores	Firms in Bottom 75% of PLi Scores	PLi Performance Premium for Highly Aligned Companies
Growth in sales	3.42°	2.91°	18%
Profit margin	3.51°	2.93°	20%
Growth in market share	3.39°	2.84°	19%
Return on investment (ROI)	3.58°	2.92°	23%
Cost reduction	3.56°	2.84°	25%

Best Practices

The remainder of this chapter will report the results of a series of field interviews conducted by the University of Tennessee and affiliated faculty of major supply chain leaders such as Caterpillar, Dell, Eastman, Ecolabs, IBM, Mondelez, and P&G. The interview results uncover best practices in purchasing and logistics integration, showing how some companies are "bending the chain."

This section also provides a helpful short list of effective leadership actions a supply chain leader can take today.

The supply chains that consistently deliver the strongest business results have the following Purchasing/Logistics characteristics:

- **BP 1**—Fully integrated, end-to-end supply chain organization with common metrics

- **BP 2**—Talented supply chain organization that rewards people for in-depth mastery *AND* end-to-end supply chain leadership

- **BP 3**—Purchasing and logistics network with an operating decision framework based on best overall total value of ownership (TVO)

- **BP 4**—Effective information systems and work processes that enable superior business results by providing their multifunctional supply chain teams the proper tools and information

Best Practice 1: Fully integrated end-to-end supply chain organization with common metrics

We have learned from decades of S&OP work that a business's *demand creation* activities are most effective when they are housed in a common organization. Similarly, *demand fulfillment* activities are most effective when they are integrated under a common supply chain organization. The best, most enduring results occur when everyone in the supply chain organization is focused on delivering superb supply chain results (customer service, quality, safety, cost, cash, and so on).

For example, large, successful global consumer goods businesses have learned (the hard way) about the vital importance of fully integrated supply chain organizations. These companies are structured with an "end-to-end/fully-integrated" supply chain organization headed by a common leader. These organizations include purchasing, logistics, operations/manufacturing, engineering, innovation management, quality, and others. The common supply chain leader drives an energizing vision, single direction, common scorecards, and consistent rewards. Thus, 100 percent of the organization is focused on meeting consumer/customer needs and delivering total value to stakeholders.

These best-in-class designs are not without challenges. Frequently, purchasing owns results beyond the supply chain, including contracts for marketing spending, indirect spending, R&D suppliers, and external contractors. This creates pressure to have an executive-level purchasing manager who reports to the CEO. One global supply chain has worked through this issue by formalizing the responsibility of the purchasing VP to the global supply chain officer.

Example

A global chemical company has likewise leveraged a partnership between corporate purchases and the supply chain. This company found it necessary to change the language and create a culture called "integrated global supply chain" to highlight the need for purchasing and supply chain teamwork. This partnership between purchasing and the rest of the supply chain ensures a common direction and reward system.

We have found that in organizations without a fully integrated end-to-end structure, the most effective first step is to develop these types of partnerships. The organization benefits from partners' common vision, direction, and rewards until a more long-term structural change can be implemented.

It is important to note that these types of leadership partnerships are, by their nature, dependent on the individuals involved. It can be expected, then, that these partnerships will vary as personnel change. Therefore, it is critical that leaders view these partnerships as transitions on the path to an organizational solution.

A second challenge involves the depth of integration.

Examples

- One Fortune 500 global supply chain leader has had an integrated end-to-end supply chain for the last three decades. The company has enjoyed improvements in cost, cash, customer service, and quality. Over the years the integration has been maintained at the top of the organization but has drifted at the category teams (middle level). Functional areas (such as purchasing and logistics) became convinced that, because of internal productivity improvements, they needed to focus on their own "primary measures." Unfortunately, these middle-level teams are where 90 percent or more of the decisions impacting cost, cash, quality, and service are made. A renewal of the original end-to-end vision at all levels of the organization is now necessary.

- A successful mid-sized company has recently implemented an integrated end-to-end supply chain design. The driving factor behind the change was the inability to deliver long-term business cost goals. After a decade of strong but independent savings work by the purchasing and logistics functions, the "well was running dry." The biggest ideas were no longer inside the departments but at the supply chain integration points across the departments (optimizing piece price versus transportation cost, optimizing piece price versus sourcing location). The most systemic solution was to form and reward a fully integrated team. The organization was delivering 2.5 percent net savings but now, after forming an integrated team, has strong action plans to deliver the business need of 4 percent net cost savings.

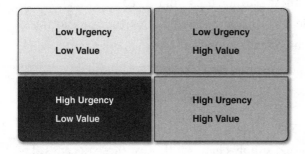

Figure 5-6 Multidiscipline Supply Teams for High-Value Work

The leadership/organizational structure is only one part of the fully integrated, end-to-end supply chain. Multidiscipline supply chain teams must be involved in strategic supplier selection and development.

You have heard the saying "Do it right the first time" your entire life. Best-in-class supply chains take this to heart. Creating the best total value supply system the first time prevents non-value-added costs, quality defects, customer service defects, and unproductive inventory while most efficiently utilizing your limited resources. This is broadly accepted but difficult to execute. Day-to-day business pressures often push managers into high-urgency/low-value activities, diverting attention from those high-value activities that can make a lasting impact.

Best-in-class supply chains utilize multidiscipline teams to manage supplier selection and develop strategic suppliers and critical materials. This ensures that the right resources are involved to develop the best end-to-end supply chain solutions. These supply chains leverage purchasing as the leader of the supplier selection/development teams. The goal is to have a clear, single point of accountability while ensuring an integrated process. These multidiscipline teams include all the relevant elements of the end-to-end supply chain (for example, engineering, logistics, manufacturing, purchasing, innovation management, quality, Six-Sigma resources, and so on). Moreover, best-in-class supply chains prioritize the level of resource involvement with the greatest business impact. Multidiscipline supply chain teams are heavily involved with the most important suppliers/materials while auxiliary teams manage less critical decisions. Additionally, many companies use senior, experienced (in multiple supply chain components) supply chain leaders in broader supplier selection teams with the expectation that they will resource experts when needed.

Examples

- A global information technology leader has a simple and transparent expectation for the use of multidiscipline supplier selection and development teams. The first time an employee does not use this kind of team, he or she receives a warning. The second time results in termination. This extreme principle is being utilized to change the culture and ensure TVO requirements are delivered.

- This same supply chain leader requires that all supplier selection teams maintain responsibility for supplier development. "The development of our supplier partners is critical to delivering our long-term goals. We want the accountability for selection and development to be consistent. Decisions in the selection process are owned through execution."

- A major retailer is linking merchandising with its supply chain resources on supplier selection. This same company is forming multidiscipline teams to work with private label supplier development teams. Additionally, a director of supplier collaboration has been appointed to drive faster progress in these areas.

- A major global CPG company has benefited from multidiscipline teams for multiple decades. These teams have facilitated TVO at a category or brand level. The opportunity is to multiply the scale, leveraging strong supplier partner capabilities across categories. The key action plan is to involve "other category" multidiscipline teams in their supplier selection/development processes to harvest scale within a supplier. An example of this is working to align common chemical specifications across categories to increase supplier scale/volume discounts.

Best Practice 2: Talented supply chain organization that rewards people for in-depth mastery and end-to-end supply chain leadership

For years, logistics and purchasing leaders have argued that these two vital elements of the supply chain must be in separate organizations with different recruiting, training, rewards, and rituals. Typical arguments included the following:

- Purchasing is an externally focused organization.

- Purchasing is commercial work, not technical work.

- Purchasing requires strong entrepreneurial skills.

- Logistics must stay focused on delivering this week.
- Logistics is busy leading inventory and customer service.
- Logistics must have a strong day-to-day team relationship with manufacturing plants.
- Logistics must be expert planners and APEC certified.

The most effective supply chain leaders have created new paradigms. The breakthrough improvements in cost, cash, and service lie in the seams of the supply chain, but integrated approaches are needed to achieve these benefits. Therefore, supply chains must strive for functional depth and the necessary end-to-end breadth of supply chain skills. In our research we found multiple examples of best-in-class supply chains that have broken through to this new paradigm.

Examples

- A large industrial equipment company is requiring its purchasing and logistics resources to come to senior business/product managers with integrated action plans and goals.
- A major global chemical company has a system to move all new supply chain managers through multiple supply chain disciplines.
- Multiple top-tier organizations are now requiring that a single senior supply chain manager with broad end-to-end skills be actively involved in upfront innovation work processes with R&D. As a result of significant corporate productivity goals, the days of sending multiple supply chain leaders are over. As one manager put it, "We must have supply chain leaders with strong purchasing and logistics skills influencing the new product/supply chain decisions of our future."

Supply chain executive VPs are requiring that purchasing and logistics leaders do more than create "great purchasing or logistics" talent. These disciplines must develop in-depth mastery to drive results for today while simultaneously building end-to-end supply chain skills to meet the complex supply chain problems and opportunities of tomorrow. The secondary benefit of these senior leadership expectations is the creation of a larger pool of future supply chain executive leadership talent.

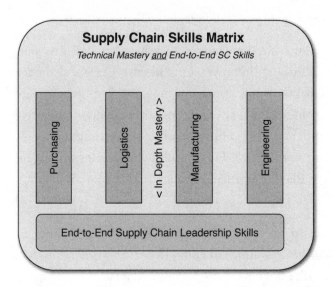

Figure 5-7 Supply Chain Skills Matrix, In-Depth Mastery, and End-to-End SC Skills

Best Practice 3: Purchasing and logistics network with an operating decision framework based on best overall total value of ownership (TVO = total cost of ownership plus level of customer creation)

Many purchasing teams have been using broad supplier scorecards in the supplier selection process for years. Nevertheless, supply chains continue suffering from inaccurate prediction of supplier cost, significant quality issues/rework costs, and capacity issues because of poor supplier reliability.

Our research has shown that the existence of a supplier scorecard is insufficient to drive excellence in supplier-driven supply chain metrics.

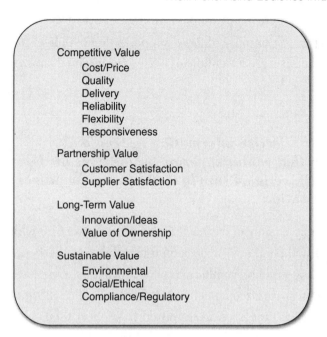

Competitive Value
  Cost/Price
  Quality
  Delivery
  Reliability
  Flexibility
  Responsiveness

Partnership Value
  Customer Satisfaction
  Supplier Satisfaction

Long-Term Value
  Innovation/Ideas
  Value of Ownership

Sustainable Value
  Environmental
  Social/Ethical
  Compliance/Regulatory

Figure 5-8 Supplier Selection Metrics

Today, most supplier selection and development are led and managed by purchasing, and decision making is largely based on piece price. However, the most effective supply chains have successfully transitioned to decisions based on TVO. This requires broader supply chain involvement (see Best Practice 1) and a commitment to TVO-based decisions.

Examples

- A world-class global information technology company uses internal supplier selection consultants to review supplier decisions. This has significantly changed the reward system. In the rapidly changing information technology business, mistakes created by narrow piece price decisions can make or break profit goals for the company.

- A global industrial equipment company found significant defects in its supplier scorecards. The purchasing teams were measuring piece price and supplier on-time delivery. Unfortunately, 90 percent or more of these suppliers do not deliver the materials. This is a great example of having the wrong measures on the scorecard.

Best Practice 4: Effective information systems and work processes that enable superior business results by providing multifunctional supply chain teams the proper tools and information

Finally, a supply chain can have a fully integrated structure with talented, well-trained people who focus on total value yet still do not deliver best-in-class supplier results. Empowered teams must have the tools to execute with excellence. Robust and efficient information systems and work processes are required to support total value creation.

We have interviewed many companies that start with a holistic supplier scorecard but struggle with placing a value on customer service/quality issues, cost of inventory, environmental incidents, reliable supply/delivery, and so on.

Some of these elements have a cumulative impact. The first environmental issue may have a limited impact, but multiple incidents can cause significant legal costs, time commitment, investments, and business disruption (in extreme cases). How do you place a cost on these types of elements? We have found that the best-in-class supply chains *partner with finance* to align on the value of these items. These can be intense debates. Our suggestion is to create a starting point and adjust as you learn.

Examples

- A medium-size service-oriented company recently changed its executive vice president of supply chain. The new executive found that the reported supplier savings were not making it to the bottom line. The reward system was based on gross purchase price savings. The EVP changed the supply chain reward system, focusing the purchasing cost measure on net savings, thus creating an immediate impact on corporate results.

- Likewise, a large global consumer goods company had to change how it valued quality investments. Historically, investments for handling new materials and suppliers were approached with a "zero capital mindset." New materials/suppliers were simply brought in, and it was up to the facility managers to utilize their extremely high-skilled workforce to develop low-cost solutions. This had a compounding effect, in which every couple of years a major quality-based capital appropriation was required. Finance and the supply chain leader aligned on including "fair share" capital as part of these types of supplier selections to realistically model the total cost.

Robust work processes are required for managing these decisions. Clear metrics, decision authority, multidiscipline team criteria, monthly reviews, and leadership involvement are a few of these important processes.

Examples

- A global chemical company has rigorous quarterly review processes (by chemical segmentation). The quarterly reviews include a full analysis of successes and failures, with action plans to drive continuous improvement. Additionally, the reviews are based on a holistic total value scorecard, including suppliers' work on innovation (supporting the chemical company's initiatives and internal supplier innovation).

- A global CPG executive vice president requires top supply chain leadership reviews of critical supplier decisions. "Left alone, the culture reverts to a purchasing process based on piece price. To change this culture, supply chain leaders must be actively involved in the reviews—pushing for total value, driving to determine the cost of quality/service issues, and incorporating the true cost of cash." The active leadership involvement and reviews are changing the culture, driving better decisions, and training the organization on what will be rewarded.

This is complicated as the world becomes more global. Many of these processes must work across different regions. We now live in a virtual world requiring virtual processes.

Example

- A mid-sized global service company has utilized multidiscipline teams. Because of the global nature of its business, these teams are virtual, with participants from around the world. Their inefficiencies come from the virtual work process. Teamwork is a key issue. The virtual team did not have the level of teamwork experienced in co-located facilities. The root causes were little/no informal time for team communication, lack of team building to build trust, and the time lag as the team collectively solved problems. Implementation of improved communication tools for virtual teams is a critical action plan.

Seven Actions a Supply Chain Leader Can Take Today

The value of the research, best practices, and examples is determined by how they can change your supply chain leadership. Following is a list of potential actions you could take today to make a difference in your organization and business results.

1. *Get it on business leader scorecards.* Work with your general managers/business leaders to ensure holistic measures are on the business/general manager scorecards. Profit and cost are consistently on these high-level scorecards, but quality, cash, and customer service may not be. Including supply chain

excellence measures on the business scorecard enables you to lead based on business priorities.

2. *Champion TVO.* It is not enough to talk about the use of total value of ownership with your direct reports. Talk the importance of total value with supplier selection and development as part of your communications (meetings, calls, printed documents, supply chain goals/action plans), participate in supplier selection and development reviews for the most strategic suppliers/materials, and ensure that the rewards for supply chain people are consistent with TVO.

3. *Make R&D your best friend.* Create a strong partnership with the research and development leader. Consider co-locating your office with the R&D leader to facilitate teamwork and symbolize a seamless technical community. The SC leader and the R&D leader should have common expectations, including active, up-front involvement in new initiative supplier decisions and product design to optimize innovation that delivers consumer, customer, supplier, community, and shareholder needs.

4. *Be clear.* Set clear expectations for use of multidiscipline teams on supplier selection. Ensure people know what process is expected for what type of suppliers. Do this publicly and in written communications. Enable your multidiscipline teams to do the work. Help your global virtual teams get the tools they need to succeed.

5. *Champion an end-to-end and integrated supply chain organization.* If your supply chain team is not end-to-end and fully integrated, create a plan to make this happen. This is not easy or straightforward leadership work in many companies. Barriers to creating your supply chain organizational vision include commercial business leaders who have other ideas, existing acquisition agreements (including personal contacts), and historical systems. Stay committed to achieving the vision, and make progress with every organizational opportunity.

Align on a common direction. If the purchasing and logistics teams have different leadership, partner with these leaders to ensure both organizations have a common supplier direction, scorecards, and rewards. This alignment can precede more complex organizational structure changes and deliver immediate business improvement. This type of clear organizational direction creates more leadership work, because the two leaders must speak with a common voice. But the investment with your partner to create this common voice will reward both of you with better decision making (until the structural change is made).

6. *Build talent focused on the end-to-end supply chain.* Create a principle for strong end-to-end supply chain skill requirements for leadership positions in each supply chain discipline. Today's business challenges require supply chain leaders who can build strong "links" in the supply systems and resolve integration problems. Disciplined leaders who have demonstrated successful results in multiple disciplines will strengthen the capability of the total supply chain leadership team.

7. *Partner with finance.* Work with finance leadership to align on how your multidiscipline teams quantify quality, customer service, environmental, sustainability, delivery, inventory, and the like. A primary leadership role is enabling the organization with clear expectations and aligned measures. Delegation of this leadership work "freezes" most teams. Create a starting point on how to value, learn, and adjust.

How High Is Your PLi?

Our research shows there is a tremendous benefit when firms align purchasing and logistics activities across the value chain to facilitate collaboration.

So, how integrated are your firm's purchasing and logistics functions? *Very*, you say? Are you sure?

Why not test this with a quick checkup that will answer the question, "How high is your PLi?" Perhaps purchasing believes there is excellent collaboration while logistics does not, or vice versa.

Send copies of this brief self-test to key members of your purchasing and logistics teams, and ask them to return them to you. See where their answers are aligned and where they are different. This is not a scientific tool but one designed to provide insight into how both groups view their level of collaboration. This eye-opening exercise could lead to valuable process improvements that can raise your PLi.

Table 5-8 How High Is Your PLi?

Answer the following questions on a 1-to-5 scale.

Apply the questions based on your business.

Scale:

5—fully implemented, producing strong results, cultural norm

3—implemented but not a cultural norm and requires leadership recruitment

1—not implemented, being discussed

Question	Score	Comments
1. Do you have a fully integrated end-to-end supply chain organization where purchasing and logistics report to the same supply chain VP?		
2. Do you have one common supply chain vision, direction, and rewards system for all purchasing and logistics personnel?		
3. Do you have a common supply chain scorecard where all disciplines in the supply chain report results?		
4. Do you measure supplier selection, development, and other operational decisions based on total value to your company?		
5. Does your organization have clear measures for the value of inventory, quality, and customer service to include in the total value equation?		

Question	Score	Comments
6. Do you utilize multifunctional teams (i.e., R&D, finance, operations, quality, engineering, logistics, purchasing) appropriate for your business to select and develop strategic suppliers and materials?		

Endnotes

1. Theodore P. Stank is the Bruce Chair of Excellence in Business, J. Paul Dittmann is the Executive Director of the Global Supply Chain Institute, Chad W. Autry is the W.J. Taylor Professor of Supply Chain Management, Michael J. Burnette is a Global Supply Chain Institute Fellow, and Daniel A. Pellathy is a doctoral candidate, all within the department of Marketing and Supply Chain Management at the University of Tennessee's Haslam College of Business. Kenneth J. Petersen is the Dean of the College of Business and Economics at Boise State University.

2. Oliver Wight International, Inc. 2000. *The Oliver Wight ABCD Checklist for Operational Excellence*, 5th ed. New York: John Wiley and Sons Inc.

6

Getting Aligned: The Benefits of Integrating Market, Environmental, Social, and Political Strategies Within the Organization

**By T. Russell Crook, Michael P. Lerman,
and Matthew C. Harris**[1]

When executives take the time to develop and implement strategy, their firms typically perform better. The alignment framework presented in this chapter provides guidance for executives regarding some key considerations in what is important in market strategy, and how executives can integrate market strategy with environmental, social, and political strategy. On the one hand, market strategies focus on customer attributes and are geared toward building advantages over competitors. On the other, nonmarket strategies, such as environmental, social, and political strategies, are primarily geared toward key stakeholders inside and outside the firm, rather than customers. In short, when executives integrate and execute market and nonmarket strategies, their firms will avoid costly mistakes and ultimately perform better.

Why do executives develop and implement strategies? It is quite simple. They want to help their firms improve performance, whether it is growing top-line sales, improving costs, or sometimes to "right the ship" and turn things around. A key goal of strategic management research is to help executives find ways to improve their firms' performance. Research indicates that firm performance is strongly driven by a firm's market position within an industry (15 percent) and whether the industry conditions are favorable or unfavorable (19 percent).[2] A key implication is that firms that take the time to find the right industry and think through their customer value proposition, such as whether they want to provide affordable prices to a broad set of consumers or appeal to a small set of well-heeled customers, can improve their performance. In short, firms that are thoughtful in how they develop and implement market strategy well outperform their rivals.

Within the past decade, executives have heightened their focus on what some have referred to as nonmarket strategy. Unlike market positioning, this facet of strategy encompasses environmental, social, and political strategy. Environmental strategy is defined as a firm's effort to reduce its impact on the natural environment. For some firms, this involves considering and investing in activities that reduce their carbon footprint through lower consumption of fossil fuels. Social strategy is a firm's effort geared toward building and sustaining relationships with key stakeholder groups. Such strategy is often geared toward activities that help the underprivileged (locally or globally) and make the world "a better place" (for example, Tom's Shoes). Political strategy refers to efforts to engage politicians and governmental entities. Political strategy might involve making campaign contributions and/or grassroots lobbying efforts. Ultimately, these "nonmarket strategies" are geared toward building bridges or buffering firms from potential problems down the road. In either case, though, recent research shows that these facets of strategy can shape firm performance, and do so in a meaningful way.

For firms to get the most from their market and nonmarket strategies, these strategies must be aligned so that the actions complement each other in a meaningful way. In this chapter, we introduce what we refer to as the "alignment framework."[3] This model provides a framework for executives to think strategically about aligning strategy.

Without alignment, executives run the risk of engaging in activities that are divorced from one another and, ultimately, do not meet their potential for their firms.

How Do Executives Engage in Strategic Management?

Many, if not most, executives have heard of strategic management. In fact, it is not entirely uncommon for many people to believe that just about every important executive action is strategic or that, because the executive is acting, he or she is practicing strategic management. This obfuscates what strategic management is and how strategies differ from actions. Strategy is about understanding the environment (external and internal) of a firm and thinking through how a firm will survive and thrive in the marketplace. Actions are simply the tactics that firms choose to implement along the way. When firms practice the ongoing process of developing a vision and mission for the organization, setting goals (for example, specific, measurable, aggressive, realistic, and time bound), developing strategy, implementing actions, and then evaluating the performance of the firm (sales growth or profit performance), they are engaging in "strategic management."

Of course, an important part of strategic management is analysis. This is where executives take stock of the environment so that sound strategies can be conceptualized, developed, evaluated, and implemented. The set of strategies pursued by a firm is typically the result of complex interactions among analyses, decisions, actions, and, ultimately, evaluating results. As shown in Figure 6-1, for most firms, especially large ones, both market strategies (brand building to increase "mind share" in the marketplace) and nonmarket strategies (political lobbying) are usually the product of considerable analysis. Putting together strategic plans, or the formulation of these strategies, often comes after data collection and interpretation regarding the industry, the firm's competitors and stakeholders, and other important factors.

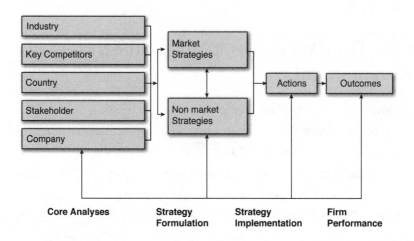

Figure 6-1 The Alignment Framework

It is tempting to gather as much data as possible to make informed strategic decisions. However, this temptation can sidetrack the actual implementation. Some executives suffer from "paralysis by analysis" and get so caught up in the analysis that either they forget to act, or they act in such a delayed fashion that the payoff is very limited. Thus, an important implication is for executives to get to the point where they believe they have conducted sufficient analysis to make thoughtful and informed strategic decisions. But, in doing so, they should recognize that there are diminishing returns to incremental analyses.

After analyses have been conducted and decisions have been made, action is needed. In other words, executives and other people within the firm then must implement the strategy. Implementation involves taking a series of actions, such as allocating resources, acquiring and/or deploying talent, and sometimes, changing organizational structures and incentive systems. Finally, executives must measure the outcomes of the strategy. Performance data needs to be collected and used to compare outcomes to the goals that were set earlier on in the process. Then this information can be used to make adjustments. This is a multistep process that requires executives to revisit the process regularly so that they ensure the firm's performance is meeting expectations.[4]

The Alignment Framework

The approach to strategic management just described is depicted in the alignment framework. By using this framework, executives can clearly see the big picture, including their firm's position, how it competes against other firms, as well as global forces, such as technological developments, macroeconomic changes, and several other factors (such as in the microenvironment) that will affect their business. Second, certain portions of the framework require executives to make predictions; this means that managers must develop one or more scenarios concerning the future of their industry—such as how competitors are likely to behave—and to derive the implications of these scenarios for today's plans and actions. Finally, by collecting and analyzing relevant data, executives can ground their decisions and recommendations in reality. Thus, the strategies that emerge from the use of the alignment framework are a starting point, but they are credible and feasible. Because of this, the strategies are more likely to succeed than strategies that are not based on data, and are purely based on "feel" and "gut."

The specific questions that guide the decision-making and data-collection processes are listed in Table 6-1. As shown, there are 22 sets of questions on which the various analyses are based. Gathering data to answer those questions is often a time-consuming and difficult process. Because of this, the analyst may have to dilute or even forgo acquiring some potentially important information to keep moving forward. However, after the analyses have been conducted, executives can have greater confidence in their strategic decisions, especially because of their current knowledge about their own firms and resources as well as their industry, competitors, stakeholders, and international markets. Armed with this knowledge, executives are in a better position to understand the competitive landscape and make recommendations about what actions their firm should take.

Table 6-1 Key Questions for Strategy Development

Industry Analysis

- How attractive is the industry, now and in the future?
- How do the threats of new entrants, substitutes, the amount of competitive rivalry, as well as the bargaining power of buyers and suppliers shape profits at the firm level?

Competitor Analysis

- Who are the key competitors in the industry, and can they be classified into subsets of groups?
- What is the nature of competition within and across the groups?
- How are firms in each of the groups likely to compete in the future?

Country Analysis

- What is the expected demand for the firm's products and services in the targeted international market?
- What are the risks (type and severity) associated with the venture?
- How should the business be launched (startup via greenfield, partnership, acquisition)?
- What are the key legal and regulatory constraints faced by the firm?
- How might legal and regulatory factors change in the future, and what opportunities might result?

Stakeholder Analysis

- Who are the firm's key stakeholders?
- What are the expectations of each stakeholder group?
- Are there existing and potential stakeholder relationships that could affect the firm?

Company Analysis

- What is the company's position in its industry?
- What is the company's existing strategy for competing in the industry? Are company resources and capabilities well positioned to capitalize on opportunities?
- How is the company organized and managed to pursue its strategy?
- What has been the company's performance in recent years? What explains this performance? How does the company's performance compare to competitors?

Key Questions for Market and Nonmarket Strategy Alignment

- What is the firm's current reputation regarding environmental concerns? Have political relationships been established? Does the company have strong social relationships?
- What are the inherent environmental risks associated with the firm's daily activities? How can legal compliance and safety be improved?
- What political actors make decisions that affect the firm, and how can relationships be established with those actors?
- What indirect stakeholders could the firm's resources benefit? How will benefiting those stakeholders contribute to the firm's desired image?
- Do the outcomes of the nonmarket strategy contribute to the market strategy goals? How can the strategies become more closely aligned?

Applying the Alignment Framework

The Alignment Framework is based on a collection of concepts and strategy tools that have stood the test of time. If executives take the time to learn more about its concepts and tools, and gather information required to answer key questions, they will have a solid picture of what is taking place in the environment and within the firm. However, merely using the framework does not guarantee that a firm will win in the market and enjoy stronger performance. To achieve that objective, managers must craft strategies that work and are aligned with one another.

For example, market strategy focuses on positioning in the market. This allows a firm to determine the proper product to develop, how to differentiate that product, and what geographic market to focus on. Before considering any of the above, however, it is important to determine if there is potential for profit in the environment in which the firm exists. To understand profit potential, an understanding—individually and in connection with each other—of the industry, competition, country, stakeholders, government, and the company itself is critical.

Industry Analysis

When applying the framework, the first step is to conduct an industry analysis. This helps executives understand how certain key environmental factors affect the firm's operations and, ultimately, performance. Conducting an industry analysis allows executives to better understand how much profit potential exists in an industry. Although many firms might be extending their reach and entering new industries (related or unrelated), according to Michael Porter, it is important to first understand whether the industry has profit potential. Doing so helps inform decisions about whether entering a new industry makes financial sense, and thus avoid making "herd mentality" decisions. In order to make an assessment regarding profit potential, Porter's work teaches us that it is essential to consider the intensity of competitive rivalry, the bargaining power of buyers and suppliers, the potential for new entrants, and the viability of substitute products or services.[5] These are known as Porter's "Five Forces," and his model has held up for more than 35 years. Understanding how these Five Forces work together helps shed light on whether profit potential exists. Following is a more detailed view of each force, the combination of which paints a comprehensive picture of the competitive environment in a given industry.

1. Intensity of Competitive Rivalry

When the intensity of competitive rivalry in an industry is high, profits for firms in the industry suffer. Essentially, firms can grow their revenue in one of two ways: they can capture new revenue from industry growth, or they can capture market share from a competitor. Rivalry increases when industry growth is low, when customers can easily switch among products or services offered by the industry (that is, low switching costs), or when firms offer similar products and services (that is, little differentiation). For growth-minded companies, when industry growth slows, the firm must then steal customers from other firms to meet growth objectives. Also, if little differentiation exists, firms must compete on price to attract customers. The more intensely firms compete on price rather than product characteristics, the lower the profits for the firms in the industry.

The grocery/retail industry has particularly intense rivalries because the primary switching costs for customers is the cost it takes to drive or walk to a different store, and many products are similar. As described earlier, this creates a situation where firms must compete on price. Walmart is a great example of a firm that created a strategy to ensure the lowest prices possible, and it accomplished its goals through establishing an economically efficient distribution process. The key implication is that high competitive rivalry often results in aggressive pricing strategies. Pricing, in turn, places a cap on the profit potential of firms in the industry.

2. Buyer Power

When the bargaining power of buyers is high, they can demand price concessions from firms in an industry. Often, when industries are characterized as "high competitive rivalry," buyers have more power relative to sellers. Buyers also have more absolute power when there are few buyers relative to sellers, or when buyers purchase large volumes. Generally, as buyer power decreases, firms can charge higher markups and improve their profitability. An example of weak buyer power is movie theater customers. When customers go to see a movie—assuming they haven't sneaked in a snack—they have no source of beverage or food other than from the movie theater concession stand. As a result, the movie theater can raise prices to an amount that a consumer would not typically pay.

Examples of industries with high buyer power are somewhat uncommon when the public sector is not the buyer. However, some examples of industries with concentrated buyer power can be found in agriculture. For example, 50 percent of all tobacco grown is purchased by three companies. Similar concentration of buyer power is found in the cocoa beans market, or in a coal mining town where the mine is the only buyer of labor.

3. Supplier Power

The bargaining power of suppliers is high when the supplying industry is more concentrated (has fewer sources) than are the buyers, when few close substitutes exist for the item offered, or when there is high product differentiation among suppliers. Intel and AMD, suppliers of computer processing chips, are examples of firms that also have strong supplier power. There are very few substitutes to their product, and thus they can control their prices and contracts more than the typical supplier. Similarly, Luxottica eyewear is involved in the production of more than 80 percent of the world's glasses. The music industry is an example of a situation where supplier power is low. The suppliers—musicians—are in great abundance, whereas the record companies that sign musicians are of limited quantity. This creates a situation where the buyers—record companies—not only get to handpick their musicians, but also are able to achieve favorable contract terms.

4. Substitutes

Firms must also consider the viability of substitutes. Whereas competitive rivalry refers to competing firms selling the *same* good or service, substitutes focus on the viability of closely related goods.

When close substitutes are available, firms must devise ways to make their products and services more attractive than the substitutes. What is important here is that a substitute can, more or less, accomplish the same objective. For example, buses, planes, trains, and cars are substitutes on relatively short-to-medium distance routes. It is important to remember, however, that a substitute is dependent upon what a specific customer is looking for. Direct substitutes to a product— items that attempt to offer the same value proposition to customers— should be most heavily considered.

5. New Entrants

Finally, the threat of new entrants refers to the prospect that new players will enter an industry. New entrants generally lead to an erosion of industry profits. The likelihood of new entry is low if an industry has high capital requirements, saturated distribution channels, large economies of scale, and restrictive government regulations. For example, the threat of new entrants in the airline industry is relatively low because it requires significant financial backing, a fleet of aircraft, access to an operating certificate, human capital, and so on in order to compete. In other words, the barriers to entry are large. In the pizza restaurant industry, however, the threat of new entrants is very high. Although entry costs are moderate, there is not significant differentiation among products, and if the initial costs can be afforded, a new restaurant can usually get an opportunity to earn market share. A key implication is that because it can be relatively easy to start a new pizza restaurant, it places a cap on what current competitors can charge. If they charge too much, a slew of new competitiveness would be unleashed among new competitors that would slash prices. Generally, the industries that exhibit higher profits also make for more appealing targets for new entrants.

Competitor Analysis

How should a firm respond to competitors' actions? In most large firms, competitive moves are determined using information on competitive intelligence. In conducting their analyses, competitor-intelligence professionals might use the concept of strategic groups, a conceptual and statistical procedure that classifies competitors into groups of firms having similar characteristics. For example, Coca-Cola and Pepsi are in the same strategic group because their product is substantially the same. In fact, many soda drinkers tend to be either "Pepsi people" or "Coke people," which is a reference to this direct competition. To be maximally effective, competitive moves need to be focused primarily on direct competitors.

Country Analysis

A firm must determine the international markets that are most receptive to its product and service offerings. A host of factors need to be taken into account for a country analysis, and data quality may range from highly reliable to unavailable. Perhaps the most common and comprehensive tool for conducting country analysis is Political, Economic, Sociocultural, and Technological (PEST). When examining where your firm currently operates or where it is considering operating in the future, it is important to understand the political, regulatory, and legal environment. It is also important to understand the economic potential of the country. Can the country's citizens afford your product or service? Will there be opportunities for growth? What risks are present? What are the key trends? How should the business be launched? Understanding the social and cultural environment is also important. Unfortunately, these are also the factors where it can be difficult to gauge the situation. Finally, the technological environment refers to the changes in communications, infrastructure, and productive capabilities within the country.

Understanding these factors will require not just a look at the CIA's World Fact Book, but a "deeper dive" that also gets locals involved. For firms in the future, it will be important to move beyond simple characterizations; Western countries, especially the United States, are known for being more individualistic and for placing high value on self-reliance. By contrast, Eastern countries, specifically China, are considered to be more collectivist, placing high value on group cooperation. These differences, along with several others, are important, but such generalizations will get people only so far. Using PEST analysis can shed further light into the environment, help create even more attractive opportunities, and also help avoid costly errors (such as GM introducing the Nova in Latin America).

Stakeholder Analysis

Organizations should also direct attention to nonmarket strategies and analyze who key stakeholders are and how they can be managed effectively. For example, the dike failure at the Tennessee Valley Authority's Kingston Fossil Plant allowed more than 5 million cubic yards of coal ash into the Emory and Clinch rivers while also covering 300 acres of land in Harriman, Tennessee.[6] Ultimately, the incident caused damage to the environment and to the individuals living around the plant, many of whom had their homes stripped from the foundation. This illustrates the importance of considering not only the direct stakeholders of a business but also the nonmarket stakeholders when creating business processes.

Company Analysis

One final, important analysis is of the firm itself. Unlike the previous analyses, which for the most part were focused on external matters, company analysis involves internal factors, such as goals along the resources and capabilities needed to attain them. The overall objective of all the firm's strategies is to match its internal features and capabilities with external conditions, as evidenced by the familiar *strengths, weaknesses, opportunities, and threats (SWOT)* analysis. Ben Greenfield is one example of many individuals who have utilized the opportunity of a globalized market to showcase their strengths to as many individuals as possible. Greenfield used to be a personal trainer and mostly worked with a single individual at a time. Thanks to the explosion of the Internet, he was able to create a website, podcast, e-books, and so on that allowed him to get more widespread access to customers for virtually the same amount of effort.

Much of the strategic management research throughout the past decade draws on a resource-based view of a firm. According to this view, strategic resources are critical to company analysis and effective strategy making. The strategic resources considered in this approach can be defined as hard-to-copy assets and abilities that enable firms to sustain higher performance levels than their peers can.

Walmart is an example of a company with an asset that is nearly impossible to copy; it has arguably the best supply chain in the world. A small example of how it uses its supply chain is the hub-and-spoke expansion method. When Walmart was in its largest growth phase, it focused on finding ideal locations for distribution centers; then it built its discount stores around that central location. This not only provided substantial savings in transportation costs, but allowed an easy transition into the grocery industry; Walmart simply upgraded its already existing discount stores.

Integrating Market and Nonmarket Strategies

Equipped with sufficient answers to Table 6-1's questions, managers can devise or modify their firm's market and nonmarket strategies. Alignment between market and nonmarket strategy is necessary to optimize performance. Otherwise, by failing to consider one strategy while implementing another, the focus toward different goals causes misalignment and ultimately harms performance. Before considering how market and nonmarket strategy can be integrated, it is important to first conceptualize each individually. Then an example of good and bad nonmarket strategies will help to illustrate the importance of alignment. Finally, examination of current research will show support for the alignment of market and nonmarket strategy.

As noted earlier, market strategy broadly relates to improving firm performance through interactions with customers, suppliers, and competitors. More specifically, this is accomplished through careful consideration of what product to sell, the geographic markets to sell the product to, and how differentiation will be achieved within a specific industry. The approach to market strategy can create creative differences within the same industry. For example, pizza restaurants typically offer different types of pizza based on the geographic location of the firm in the United States. California-style pizza is known for being thin and using a variety of nontraditional cheeses. Chicago-style, in contrast, is popular for its deep-dish, thick-crust approach with mozzarella cheese. This creates an interesting competitive dynamic in the pizza business, where firms must strongly consider the specific client base they want to sell to.

Firms can also use market strategy to create a new market. By applying a rich blend of psychological and statistical analyses to customer-purchase data, grocery stores are able to anticipate trends and buying patterns. They use this ability to print personalized coupons relating to goods that they expect a specific customer to be interested in. Thus, the grocery stores that have applied this technique have been able to make sales that probably would not have occurred otherwise.

In both illustrations in the two preceding paragraphs, abundant financial implications occur as a result of the executed market strategy. This intuitively makes sense: the goal of any business is to survive by making money, and thus market strategy aims to optimize income through effective relationship building with its customers and suppliers, while outperforming its competitors. Nonmarket strategy, when well designed and integrated with market strategy, can not only affect profitability, but shift the Five Forces that shape the competitive environment of an industry. However, alignment of market and nonmarket strategies is essential. Before actively discussing this connection, it is important to further define nonmarket strategies.

Nonmarket strategies consist of environmental, social, and political elements. These elements address the concerns of stakeholder groups other than customers: employees, special-interest groups, shareholders, and governments. Nonmarket stakeholder groups are unique from customers because they do not directly engage with the firm, but they are similar in that they are still affected by the firm's actions. The stakeholders' response to firm actions plays a significant role in the success or failure of the firm. Thus, a strong understanding of the environmental, social, and political elements of nonmarket strategy can add significant value to a firm.

Environmental strategy relates to a firm's impact on the environment. Some companies, as a core of their business, deal heavily with environmental concerns. For example, the oil industry must put significant internal controls in place to ensure that an oil spill will not occur. When these controls fail, the results are catastrophic: BP's spill in April 2010 allowed 134 million gallons of crude oil into the ocean, leading to "oil-soaked birds, fish and turtles washing up on shore along the coast."[7] Despite hundreds of millions spent on developing its market strategy, this example illustrates the impact that a lack of environmental strategy had on BP.

In contrast, firms that sell outdoor equipment have incentives to ensure that the outdoors are viable areas for recreation for years to come. Patagonia Corporation ($540 million annual sales) pledges that three quarters of its materials are "environmentally sound." Patagonia also tithes 1 percent of its profits to preservation groups and grants employees paid leave to pursue environmental volunteer work. It also buys wool only from farmers (in Patagonia) who use sustainable grazing practices. As Patagonia sells to individuals with presumably some interest in enjoying the great outdoors, these activities can increase individuals' willingness to pay for Patagonia products, relative to rival or substitute goods.

Social strategy focuses on relationships with specific stakeholder groups at the local or global level. Social strategy can be particularly relevant for multinational corporations that have a production facility located in a developing country. In the case of FIJI, for example, five neighboring villages represented the primary workforce for FIJI's bottled water production plant. Thus, FIJI attempted to build a strong relationship with the indigenous population by constructing kindergarten classrooms in each village to support education, and by developing a trust fund of FJ$275,000 to support a potable water supply.[8] This helped FIJI access the precious water resources that, from a market standpoint, created differentiation and profit potential for the firm.

Similarly, Volkswagen AG has developed a reputation for paying higher than normal wages in developing markets (2.21 times the minimum wage in Brazil for starting factory workers) and fostering development in the communities where it operates. Volkswagen partners with Xinjiang Medical University in China to form an exchange program for young doctors. Because VW sold 3.27 million vehicles in China in 2013, there is likely some conscious effort to strategically integrate its nonmarket and market strategies.

Political strategy involves interactions with politicians and government entities through campaign contributions, lobbying, and grassroots campaigns. In return, firms hope to achieve access to key lawmakers and input regarding new legislation. At a minimum, firms hope to be involved in "conversations" that might affect their industry. For example, after announcing an attempt to merge with Time Warner Cable, Comcast paid a combined $114,175 to two senators from Pennsylvania. In return, Comcast earned a joint letter to the FCC Chair Tom Wheeler requesting a quickly approved merger.[9]

A key implication is that having political access can help firms achieve their goals.

Although market and nonmarket strategies have both been shown to have an impact on business operations, it remains to be illustrated how they tie together. BP's oil spill not only lost the company a significant amount of product, but also caused reputational damage that may have permanently lost BP customers. When FIJI made the market decision to base its production facility in Fiji, the nonmarket strategy of connecting with the indigenous population enabled a more productive workforce, thus increasing profitability. Last, through nonmarket political lobbying strategies, Comcast increased the likelihood of achieving a valuable market opportunity. For a deeper understanding of the relationship between market and nonmarket strategy, an example of a company effectively using market and nonmarket strategy—Walmart—and a company poorly implementing market and nonmarket strategy—Pepsi—follows.

Walmart used alignment of market and nonmarket strategy to make substantial improvements to its business. In the early 2000s, Walmart faced heavy criticism from a nonmarket strategy perspective. In 2003 and 2004, Kroger supermarkets and others forced their unions to accept lower wages to better compete with Walmart. This led to union attacks on Walmart, which was ultimately criticized in regard to low wages, inadequate health care, hurting small business, destroying the environment, and violating labor laws in overseas factories. Many considered these events to have a direct impact on Walmart's stock price, which had fallen by 27 percent since its new CEO, H. Lee Scott, took over in 2000.

To address these concerns, Walmart took several steps. First, it hired former presidential advisors Michael Deaver and Leslie Dach, as well as a public relations firm, Edelman. Then Scott began meeting with critics, such as the Investor Responsibility Research Center (IRRC) and others, working with their leaders to come up with solutions. Walmart paid for a full-page advertisement in 100 newspapers across the nation in early 2005, which gave facts about its employee wages and benefits while strongly refuting the critics of the organization. It conducted an internal audit that revealed employees who began working at Walmart became less likely to stay on government assistance. Then, in 2005, Walmart devoted $35 million to a 10-year plan that would create 140,000 acres of land available for hunting, fishing, and other outdoor activities.

Walmart also made its products more environmentally sound. For example, Walmart worked with liquid laundry detergent suppliers to take water out of the detergent. Doing so enabled a more concentrated detergent that also comes in a smaller bottle, allowing for reduced waste on each production, without sacrificing the number of loads a bottle can do. This approach reduced the effect on the environment while also supporting a profitable product. The steps just detailed are only a portion of the steps Walmart took, but it is clear that it invested its money in wise nonmarket strategy endeavors, which directly addressed the concerns of the public. The effect can be summed by the following RT Strategies survey finding: 71 percent of Americans believe Walmart is good for consumers while 63 percent of union households hold the same belief.[10]

Pepsi is an example of a company that attempted a nonmarket strategy but did not properly align it with its market strategy. Indra Nooyi, Pepsi's CEO, made a concerted effort to create healthier Pepsi products in response to pressure from the public. This strategy included acquiring Tropicana and Quaker Oats while also creating Pepsi Next, which has fewer calories than Pepsi. Nooyi even hired a former World Health Organization official to ensure a smooth transition. For her efforts, Nooyi earned two honorary degrees and was named the most powerful woman in business five years in a row by *Fortune* magazine.[11]

Although the nonmarket strategy addressed public concerns, it had not been properly aligned with market strategy. Consumers who were drinking Pepsi wanted the product regardless of its health concerns. Thus, investors viewed the socially responsible strategy as ineffective, causing Pepsi to lose its number two position in the market to Diet Coke in 2010. Ultimately, Pepsi was forced to acknowledge failure of its plan, and it announced that the company would be moving in a new direction.

Although the two preceding examples of firms engaging in a mixed market and nonmarket strategy suggest that it is important to understand and consider both in regard to strategy formulation, research has drawn similar conclusions. Orlitzky and colleagues combined evidence from 52 studies regarding the relationship of corporate social/environmental performance (CSP) and corporate financial performance (CFP). The results suggest that social and environmental responsibility contribute to CFP.[12] Further, managers should learn to use CSP to increase positive reputation because reputation is the best way to see returns from CSP actions. In short, nonmarket strategy can be used, in part, to differentiate a firm.

A recent study investigated the relationship between corporate environmental performance (CEP) and CFP. It focused on *when* it pays to be green by looking for differences in when environmental strategy is reactive or proactive.[13] A reactive approach focuses on complying with legislation, such as proper storage of harmful chemicals or elimination of waste. Despite this, the approach generally does not involve top management, nor significant employee environmental training. Thus, the company reacts when the issue arises. The proactive approach intertwines the environmental concerns into its market strategy and often involves fine-tuning of the business processes to prevent environmental issues from ever occurring. Surprisingly, the results suggest that firms cannot expect a significantly higher CFP if they are proactive rather than reactive. This suggests that the most important criteria are simply to comply with legislation. Interestingly, the study does suggest that U.S. firms benefit more than international firms as a result of CEP. Fowler and his colleagues posit that this is due to strict legislation, where firms that do not comply are punished. Based on the results of the study, there is compelling evidence that regardless of the firm situation—small, large, private, public, and so on—there is a financial benefit to be taken, so long as market and nonmarket strategies are aligned.

Actions and Performance

Successful market and nonmarket strategies require executives to delegate responsibility for implementation to others. Even well thought out and integrated strategies can fall apart at this stage. Because most new strategies have cross-functional implications, they therefore have the potential for disrupting various areas of the firm. For example, ineffective strategy implementation turned Firestone, which was leading the U.S. tire industry at the beginning of the 1970s, into a firm that saw its market share eroded and ultimately acquired by Bridgestone, a Japanese firm.

Firestone faced a challenge from Michelin, which introduced radial tires to the U.S. market. The new tires were known for safety, longevity, and affordability, and thus were superior to Firestone tires. In response, Firestone, which had performed due diligence in figuring out that the radial tires were going to succeed, immediately invested about $400 million to build a new plant for radial tire production while also coordinating other factories toward this effort. Unfortunately, it did not redesign its production process to ensure proper safety standards and did not immediately stop production of the now-obsolete tires. These actions led to Firestone's plants operating at 56 percent capacity, several product recalls, and the company's ultimate failure. Firestone properly conducted a core analysis and developed a strategy in accordance with its findings, but failed to consider how the new strategy would affect the different functions of the business.

Conclusion

The main purpose of this chapter was to introduce and describe the alignment framework model as a means of improving a firm's strategic decision making and performance. We hope that the ideas discussed here lead more managers to think strategically about their firms and to develop sustainable competitive approaches.

Endnotes

1. T. Russell Crook is an Associate Professor of Management and Fancher Faculty Research Fellow; Michael P. Lerman is a doctoral candidate in Management; and Matthew C. Harris is an Assistant Professor of Economics, all of the University of Tennessee's Haslam College of Business.

2. Short, J.C., D.J. Ketchen, T.B. Palmer, and G.T. Hult. 2007. Firm, Strategic Group, and Industry Influences on Performance. *Strategic Management Journal* 28(2): 147–167.

3. The framework draws on my earlier work. See Crook, T. R., D.J. Ketchen, and C.C. Snow. 2003. Competitive Edge: A Strategic Management Model. *Cornell Hotel and Restaurant Administration Quarterly*, 44: 44–55.

4. Jenster, P., and D. Hussey. 2001. *Company Analysis: Determining Strategic Capability*. Chichester, NJ: John Wiley & Sons.

5. Porter. M. 1980. *Competitive Strategy*. New York: Free Press.

6. A Dike Failure at TVA's Kingston Fossil Plant Led to the Largest Spill in History. December 23, 2013. Available at http://www.usatoday.com/story/news/nation/2013/12/22/coal-ash-spill/4143995/.

7. BP Oil Spill: Five Years After 'Worst Environmental Disaster' in US History, How Bad Was It Really? April 20, 2015. Available at http://www.telegraph.co.uk/news/worldnews/northamerica/usa/11546654/BP-oil-spill-Five-years-after-worst-environmental-disaster-in-US-history-how-bad-was-it-really.html.

8. McMaster, J., and J. Nowak. 2009. "FIJI Water and Corporate Social Responsibility—Green Makeover or 'Greenwashing'?" *Richard Ivey School of Business*, 18.

9. Here's What $184K in Campaign Contributions Gets Comcast—A Letter of Support from Two Senators. December 11, 2014. Available at http://consumerist.com/2014/12/11/heres-what-184k-in-campaign-contributions-gets-comcast-a-letter-of-support-from-two-senators/.

10. Baron, D. 2006. "Wal-Mart: Nonmarket Pressure and Reputation," *Stanford*, 4.

11. When Corporations Fail at Doing Good. August 29, 2013. Available at http://www.newyorker.com/business/currency/when-corporations-fail-at-doing-good.

12. Orlitzky, M., F.L. Schmidt, and S.L. Rynes. 2003. Corporate Social and Financial Performance: A Meta-analysis. *Organization Studies* 24(3): 403–441.

13. Dixon-Fowler, H.R., D.J. Slater, J.L. Johnson, A.E. Ellstrand, and A.M. Romi. 2013. Beyond "Does It Pay to Be Green?" A Meta-analysis of Moderators of the CEP-CFP Relationship. *Journal of Business Ethics*. 112: 353–366.

7

Achieving Demand and Supply Integration

By Mark A. Moon[1]

One of the companies that participated in some of the DSI/Fore-casting Audit research at the University of Tennessee was a company in the apparel industry. This company, a manufacturer and marketer of branded casual clothing, had very large retail customers that contributed a large percentage of overall revenue. Understandably, it was very important to the success of this company that these large retail customers are kept in stock. If these retailers' orders could not be filled, out-of-stock conditions would result, with not only lost sales as the consequence, but potential financial penalties for failure to satisfy these retailers' stringent fill-rate expectations.

As was the case for many companies in this industry over the past decade, considerable manufacturing capacity had been offshored to sewing operations in Asia. This strategy helped to keep unit costs down, but it also had a negative impact on responsiveness and flexibility. At the time of the UT audit, the research team heard about a communication disconnect between the supply chain and the sales organizations at this company. A variety of problems had left the company with significant capacity shortages. Although these problems were solvable in the long run, in the short term, it experienced significant fill-rate problems with some of its largest, most important retail customers. Some of the most popular sizes and styles of clothing were in short supply, and customers were not happy. Supply chain personnel were working hard to address these problems, but in the short term, there was little to be done. While these supply chain problems were impacting its largest, most important customers, personnel from the field sales organization were being incentivized to open new channels of distribution and locate new customers to carry their brands. As one supply chain executive told this story, she said in exasperation, "We're out of stock at Walmart, and they're signing up new customers! What the hell is going on here?"

This example is a classic illustration of what can happen when *demand and supply integration*, or DSI, is not a part of the fabric of an organization. In this chapter, we explore the essence of DSI, distinguish it from sales and operations planning (S&OP), articulate from a strategic perspective what it is designed to accomplish, describe some typical aberrations from the "ideal state" of practice, and describe some characteristics of successful DSI implementations.

The Idea Behind DSI

Demand and supply integration, or DSI, when implemented effectively, is a *single process* to engage *all functions* in creating *aligned, forward-looking plans* and to *make decisions* that will optimize resources and achieve a balanced set of organizational goals. Several phrases in the preceding sentence deserve further elaboration. First, DSI is a *single process*. The idea is that DSI is a "super-process," containing a number of "subprocesses," that are highly coordinated to achieve an overall aligned business plan. These subprocesses include demand planning, inventory planning, supply planning, and financial planning. Second, it is a process that engages *all functions*. The primary functions that must be engaged for DSI to work effectively are Sales, Marketing, Supply Chain, Finance, and Senior Leadership. Without active, committed engagement from each of the functional areas, the strategic goals behind DSI cannot be achieved. Third, it is designed to be a process that creates *aligned, forward-looking plans* and *makes decisions*. Unfortunately, when DSI is not implemented well, it often consists of "post-mortems," or discussions of "why we didn't make our numbers last month." The ultimate goal of DSI is business planning; in other words, what steps will we as an organization take *in the future* to achieve our goals?

Our research has shown that three important elements must be in place for DSI to operate effectively: *culture, process,* and *tools.* An organization's culture must be focused on transparency, collaboration, and commitment to organizationwide goals. Processes must be clearly articulated, documented, and followed to ensure that all planning steps are completed. And effective tools, normally thought of as information technology tools, are needed to provide the right information at the right time to the right people.

How DSI Is Different from S&OP

Many authors have, over the past 20+ years, written about sales and operations planning (S&OP), and to some, this description of DSI might sound like little more than a rebranding of S&OP. Unfortunately, through many years of S&OP process implementations at many companies, S&OP has a bit of a "bad name." In our observation of dozens of S&OP implementations, we've seen several common implementation problems that have contributed to a sense of frustration with the effectiveness of these processes.

First, S&OP processes are often tactical in nature. They often focus on balancing demand with supply in the short run, and turn into exercises in flexing the supply chain, either up or down, to respond to sudden and unexpected changes in demand. The planning horizon often fails to extend beyond the current fiscal quarter. With such a tactical focus, the firm can miss out on the chance to make strategic decisions about both supply capability and demand generation that extend further into the future, which can position the firm to be pro-active about pursuing market opportunities.

Second, S&OP process implementation is often initiated and managed by a firm's supply chain organization. In our experience, these business-planning processes are put into place because supply chain executives are "blamed" for failure to meet customer demand in a cost-effective way. Inventory piles up, expediting costs grow out of control, and fill rates decline, causing attention to be focused on the supply chain organization, which immediately points at the "poor fore-casts" that come out of sales and marketing. The CEO gets excited, S&OP is hailed as the way to get demand and supply in balance, and the senior supply chain executive is tasked with putting this process in place. Where the disconnect often takes place, however, is with lack of engagement from the sales and marketing functions in the organization—the owners of customers and the drivers of demand. Nothing can make S&OP processes fail any faster than having sales and marketing be nonparticipants. In more than one company we've worked with, people describe S&OP as "&OP"—meaning that "sales" is not involved.

Third, the very name "sales and operations planning" carries with it a tactical aura. As I will argue in an upcoming section, many more functions besides S&OP must be involved for effective business plan-ning to take place. Without engagement from marketing, logistics, procurement, and particularly finance and senior leadership, these attempts at integrated business planning are doomed to being highly tactical and ultimately disappointing.

Thus, although the goals of S&OP are not incompatible with the goals of DSI, the execution of S&OP often falls short. Perhaps a new branding campaign is indeed needed, because in many companies, S&OP carries with it the baggage of failed implementations. Demand and supply integration is an alternative label, and a new opportunity to achieve integrated, strategic business planning.

Signals That Demand and Supply Are Not Effectively Integrated

As our research team has worked with dozens of companies over the past 15 years, we have witnessed many instances where demand and supply are not effectively integrated. Commonly, our team is called in to diagnose problems with the forecasting and business planning processes at companies because some important performance metric—often inventory turns, carrying costs, expedited freight costs, or fill rates—has fallen below targeted levels. After we arrive onsite, we frequently hear about problems like those in the following list. Ask yourself whether any of these situations apply to your company:

- Does manufacturing complain that sales overstates demand forecasts, doesn't sell the product, and then the supply chain gets blamed for too much inventory?

- Does the sales team complain that manufacturing can't deliver on its production commitments and it's hurting sales?

- Does manufacturing complain that the sales team doesn't let them know when new product introductions should be scheduled, and then they complain about missed customer commitments?

- Does the sales team initiate promotional events to achieve end-of-quarter goals but fail to coordinate those promotional activities with the supply chain?

- Does the business not take advantage of global supply capabilities to profitably satisfy regionally?

- Are raw material purchases out of alignment with either production needs or demand requirements?

- Does the business team adequately identify potential risks and opportunities well ahead of time? Are alternatives discussed and trade-offs analyzed? Are forward actions taken to reduce risk and meet goals? Or are surprises the order of the day?

If these are common occurrences at your company, it may be the case that your demand and supply integration processes may not be living up to their potential. It could also be the case that one or more of the critical subprocesses that underlie the DSI "super-process" may be suffering from inadequate design or poor execution.

The Ideal Picture of Demand and Supply Integration

In Figure 7-1, the circles represent functional areas of the firm, the rectangle represents the super-process of DSI, the lighter arrows leading into the DSI process represent inputs to the process, and the darker arrows leading out of DSI represent outputs of the process.

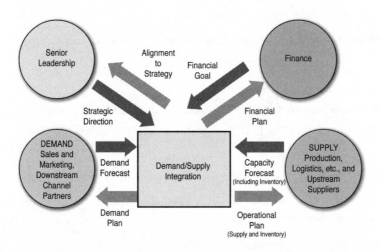

Figure 7-1 Demand/Supply Integration: The Ideal State

It all begins with the two red arrows labeled "Demand Forecast" and "Capacity Forecast." As will be clearly articulated in future chapters, the demand forecast is the firm's best "guess" about what customer demand will consist of in future time periods. It should be emphasized that this is indeed a guess. Short of having a magic crystal ball, there is uncertainty around this estimate of future demand. Of course, the further into the future we are estimating demand, the more uncertainty will exist. Similarly the "Capacity Forecast" represents the best "guess" about what future supply capability will be. Just as is the case with the demand forecast, uncertainty exists around any estimate of supply capability. Raw material or component part availability, labor availability, machine efficiency, and other supply chain variables introduce uncertainty into estimates of future capacity levels.

We'll begin with a simple example as a way of explaining how the DSI process needs to work. Assume that the "demand" side of the business—typically sales and marketing, with possible input from channel partners—goes through an exercise in demand forecasting and concludes that three months from the present date, demand will consist of 10,000 units of a particular product. We'll further assume that this demand forecast is reasonably accurate. (I know, that may be an unrealistic assumption, but let's assume it regardless!) Now, concurrently, the "supply" side of the business—operations, logistics, procurement, along with input from suppliers—conducts a capacity forecast and concludes that three months from the present date, supply capability will consist of 7,500 units. It should be noted that this outcome is far from atypical. *The fact is that demand and supply are usually NOT in balance.* So there is more demand than there is supply. The question is, *what should the firm do?*

There are several alternatives:

- *Dampen demand.* This could be achieved in a number of ways. For example, the forecasted level of demand assumes a certain price point, a certain level of advertising and promotional support, a certain number of salespeople who are selling the product with certain incentives to do so, a certain level of distribution, and so forth. Any of these demand drivers could be adjusted in an effort to bring demand into balance with supply.

Thus, some combination of a price increase or a reduction in promotional activity could dampen demand to bring it in line with supply.

- *Increase capacity.* Just as the demand forecast carries with it certain assumptions, so too does the capacity forecast. It's possible that capacity could be increased through adding additional shifts, outsourcing production, acquiring additional sources of raw materials or components, speeding up throughput, and so forth.

- *Build inventory.* Often, it is the case that in some months, capacity will exceed demand, whereas in other months, demand will exceed capacity. Rather than tweaking either demand or supply on a month-by-month basis, the firm could decide to allow some inventory to accumulate during excess capacity months, which would then be drawn down during excess demand months.

These are all worthy alternatives to solving the "demand is greater than capacity" problem. The question, then, is which of these worthy alternatives is the best alternative to solve the short-term problem while achieving a variety of other goals?

The answer is, "it depends." It depends on the costs of each alternative. It depends on the strategic desirability of each alternative. Because each situation will be unique, with different possible alternatives that carry with them different cost and strategic profiles, it is necessary to put these available alternatives in front of knowledgeable decision makers who can determine which is the best course of action. That is the purpose of the rectangle labeled in Figure 7-1 as "Demand/Supply Integration." The financial implications of each alternative are captured in the arrow labeled "Financial Goal." The strategic implications of each alternative are captured in the arrow labeled "Strategic Direction." All these pieces of information from all these different sources—the Demand Forecast, the Capacity Forecast, the Financial

Goal, and Strategic Direction—must be considered to make the best possible decisions about what to do when demand and supply are not in balance.

This simple example could be turned in the other direction. Suppose that the demand forecast for three months hence is 10,000 units, and that the capacity forecast for that same time period is 15,000 units. Now, the firm is faced with the mirror image of the first situation. Now, instead of dampening demand with price increases or reduced promotional support, the firm can increase demand with price reductions or additional promotional support. Instead of increasing production with additional shifts or outsourced manufacturing, the firm can reduce production with fewer shifts or by taking capacity down for preventive maintenance. Instead of drawing down inventory, the firm can build inventory. Again, the answer to the question of "what should we do?" is "it depends." The correct answer is a complex consideration of costs and strategic implications of each alternative. The right people need to gather with the right information available to them to make the best possible decision. Once again, DSI.

To further illustrate this "ideal state" of DSI, consider a different example. This time, assume that the demand forecast for three months hence and the capacity forecast for three months hence are both 10,000 units. (I know, that's unlikely, but let's assume it anyway.) Further, assume that if the firm sells those 10,000 units three months hence, the firm will come up short of its financial goals, and the investment community will hammer the stock. Now what? Now, both demand-side and supply-side levers must be pulled. Demand must be increased by changing the assumptions that underlie the demand forecast. Prices could be lowered; promotional activity could be accelerated; new distribution could be opened; new salespeople could be hired. Which is optimal? Well, it depends. Simultaneously, supply must be increased to meet this increased demand. Extra shifts could be added; production could be outsourced; throughput could be increased. Which is optimal? Well, it depends. Again, the right people with the right information need to gather to consider the alternatives. Again, DSI.

So we have covered the inputs to the DSI process. An unconstrained forecast of actual demand is matched up against the forecasted capacity to deliver products or services. Within the process labeled "Demand/Supply Integration" are meetings where decisions are made—decisions about how to bring demand and supply into balance, both in a tactical, short-term context *and* in a strategic, long-term context. Financial implications of the alternatives are provided from finance, and strategic direction is provided by senior leadership. However, Figure 7-1 also contains arrows that designate outputs from the DSI process. These outputs should be seen as *business plans*. There are three categories of business plans that result from the DSI process. First are *demand plans*. These represent the decisions that emerge from the DSI process that will affect sales and marketing. If prices need to be adjusted to bring demand into balance with supply, then sales and marketing need to execute those price changes. If additional promotional activity needs to be undertaken to increase demand, then sales and marketing need to execute those promotions. If new product introductions need to be accelerated (or delayed), then those marching orders need to be delivered to the responsible parties in sales and marketing. The vignette from the beginning of the chapter represented a disconnect associated with communicating and executing these demand plans.

The second output from the DSI process is *operational plans*. These operational plans represent the decisions from the DSI process that will affect the supply chain. Examples of these operational plans are production schedules, inventory planning guidelines, signals to procurement that drive orders for raw materials and component parts, signals to transportation planning that drive orders for both inbound and outbound logistics requirements, and the dozens and dozens of other tactical and strategic activities that need to be executed to deliver goods and services to customers.

Next are *financial plans*. These represent signals back into the financial planning processes of the firm, based on anticipated revenue and cost figures that are agreed to in the DSI process. Whether it is financial reporting to the investment community or acquisition of working capital to finance ongoing operations, the financial arm of the enterprise has executable activities that are dependent upon the decisions made in the DSI process about how demand and supply will be balanced. Last come those signals back to the senior leadership of the firm that the decisions that have been reached are in alignment with the strategic direction of the firm. These signals are typically delivered during the Executive DSI meetings, which are corporate leadership sessions where senior leaders are briefed on both short- and long-term business projections.

Thus, in its ideal state, DSI is a business planning process that takes in information about demand in the marketplace, supply capabilities, financial goals, and the strategic direction of the firm, and *makes clear decisions about what to do in the future.*

DSI Across the Supply Chain

Up until now, we have talked about the need to integrate demand and supply within a single enterprise. In other words, how can insights about demand levels that may be housed in sales or marketing be shared with those who need to plan the supply chain? DSI processes are the answer. But the ideal state of DSI doesn't need to be limited to information sharing within a single enterprise. Figure 7-2 represents a vision of how DSI can be expanded to encompass an entire supply chain.

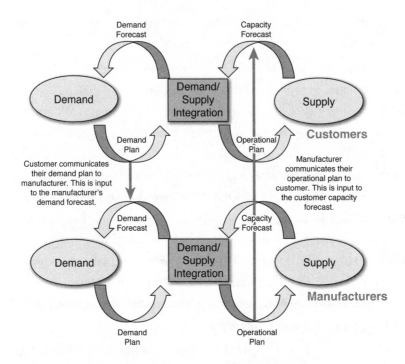

Figure 7-2 Demand/Supply Integration Across the Supply Chain

Figure 7-2's representation of DSI is a simplified version of Figure 7-1 depicted in a simplified supply chain. The red arrows show the possibilities for collaboration. First, consider the red arrow that leads from the customer's *Demand Plan* to the manufacturer's *Demand Forecast.* Imagine, for example, that the "customer" in Figure 7-2 is a computer company, and the "manufacturer" is a company that produces microprocessors for the computer industry. The computer company's demand plan will include various promotional activities that it plans to execute in future time periods to take advantage of market opportunities. The manufacturer—the microprocessor company—would benefit from knowing about these promotional activities, because it could then be able to anticipate increases in demand from this customer. Such knowledge would be incorporated into the demand forecast for the microprocessor company.

Next, consider the red arrow that points from the manufacturer's *Operational Plan* to the customer's *Capacity Forecast*. When the microprocessor company completes its DSI process, one output is an operational plan that articulates the quantity of a particular microprocessor that it intends to manufacture in future time periods. The customer—the computer company—would benefit from knowing this anticipated manufacturing quantity, particularly if it means that the microprocessor company will not be able to provide as much product as the computer company would like to have. Such a shortage would need to be a part of the computer company's capacity forecast, because this shortage will influence the results of the DSI process at the computer company. Thus, the outputs of the DSI processes at one level of the supply chain can, and should, become part of the inputs to the DSI process at other levels of the supply chain.

This same logic would apply if the "customer" was a retailer and the "manufacturer" was a company that sold its products through retail. The retailer's promotional activity, as articulated in the retailer's demand plan, is critical input to the manufacturer's demand forecast. And the manufacturer's projected build schedule, as articulated in the manufacturer's operational plan, is critical input to the retailer's capacity forecast. Companies use a variety of mechanisms to support this level of collaboration across the supply chain. These mechanisms can be as simple as a formal forecast being transmitted from the "customer" to the "manufacturer" on a regular basis. Or, the mechanism can be much more formalized, and can conform to the Collaborative Planning, Forecasting, and Replenishment (CPFR) protocol as articulated by Voluntary Interindustry Commerce Solutions, or VICS. Regardless of how this collaboration is executed, the potential exists for significant enhancements to overall supply chain effectiveness when DSI processes are implemented across multiple levels of the supply chain.

Typical DSI Aberrations

The "ideal states" of DSI as depicted in both Figures 7-1 and 7-2 are just that—ideal states. Unfortunately, a variety of forces often result in actual practice being far removed from ideal practice. I have observed a variety of "aberrations" to the ideal states articulated earlier, and three are so common that they are worth noting.

The first, and perhaps most insidious, of these DSI aberrations is depicted in Figure 7-3. Again, the figure is simplified to highlight the aberration, which is known as "Plan-Driven Forecasting." Previously in this chapter, I discussed the not-uncommon scenario where forecasted demand fails to reach the financial goals of the firm. In the ideal state of DSI, that financial goal is one of the inputs to the DSI process, where decisions are made about how demand (and if necessary, supply) should be enhanced to achieve the financial goals of the firm. In a plan-driven forecasting environment, however, the financial goal does not lead to the DSI process. Rather, it leads to the demand forecast. In other words, rather than engaging in a productive discussion about how to enhance demand, the message is sent to the demand planners that the "right answer" is to raise the forecast so that it corresponds to the financial goal. This message can be simple and direct—"raise the forecast by 10 percent"—or it can be subtle— "the demand planners know that their forecast had better show that we make our goals." Either way, this aberration is insidious.

Plan-driven forecasting is insidious because it results in a forecasting process that loses its integrity. If downstream users of the forecast—those who are making marketplace, supply chain, financial, and strategic decisions—know that the forecast is simply a restatement of the financial goals of the firm, and not an effort to predict real demand from customers, then those users will stop using the forecast to drive their decisions. I have observed two outcomes from plan-driven forecasting. First, supply chain planners go ahead and manufacture products that correspond to the artificially inflated forecast, and the result is excess, and potentially obsolete, inventory. Second, the supply chain planners say to themselves, "I know darn well that this forecast is a made-up number and doesn't represent reality. And since I own the inventory that will be generated by overproducing, I'm just going to ignore the forecast and do what I think makes sense." Here, the result is misalignment with the demand side of the company. In both cases, plan-driven forecasting results in a culture where the process loses integrity, and forecast users stop believing what forecasters say.

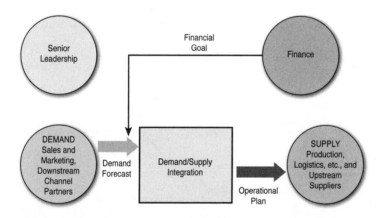

Figure 7-3 Typical DSI Aberration: Plan-Driven Forecasting

The second DSI aberration is depicted in Figure 7-4, and it is what I call "DSI as a tactical process." In this scenario, it is the responsibility of the demand side of the enterprise to come up with a demand forecast, which is then "tossed over the transom" to the supply side of the enterprise, which then either makes its plans based on the forecast, or not. Often, there really is no DSI process in place—no scheduled meetings where demand-side representatives and supply-side representatives interact to discuss issues or constraints. When this aberration is in place, there is significant risk for major disconnects between sales and marketing on the one hand and supply chain on the other. Without the information-sharing forum that a robust DSI process provides, both sides of the enterprise usually develop a sense of distrust, neither understanding nor appreciating the constraints faced by the other. In addition to the siloed culture that results from this aberration, the lack of engagement from either senior leadership or finance makes this approach to DSI extremely tactical. Often, the forecasting and planning horizons are very short, and opportunities that may be available to grow the business may be sacrificed because demand and supply are not examined from a strategic perspective. In this scenario, it is nearly impossible for DSI to be a process that "runs the business." Instead, it is limited to a process that "runs the supply chain," and engagement from sales and marketing leadership becomes very challenging.

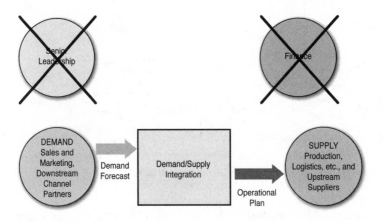

Figure 7-4 Typical DSI Aberration: DSI as a Tactical Process

The final common aberration to the ideal state of DSI is depicted in Figure 7-5. In this situation, there is little, if any, communication back to the demand side of the enterprise concerning the decisions made in the DSI process. This aberration is most problematic when there are capacity constraints in force, resulting in product shortages or allocations. Recall the scenario described at the beginning of this chapter, where the sales organization at the apparel company was incentivized to sign up new accounts while production problems were affecting deliveries to the firm's largest, most important current customer. As we have previously illustrated, the discussions that occur in the DSI process often revolve around what actions should be taken when demand is greater than supply. However, if there is no effective feedback loop that communicates these decisions back to the sales and marketing teams, then execution is not aligned with strategy, and bad outcomes often occur.

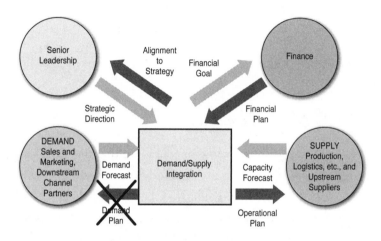

Figure 7-5 Typical DSI Aberration: Lack of Alignment with Sales and Marketing

The aberrations described here are examples of typical problems faced by companies when their DSI processes are not executed properly. Many times, aberrations like these exist, even when the formal process design is one where these aberrations would be avoided. However, siloed cultures, misaligned reward systems, lack of training, and inadequate information systems can all conspire to undermine these process designs. The reader is encouraged to look carefully at Figure 7-1, which represents the ideal state of DSI, and carefully think through each of the input and output arrows shown in the figure. Wherever an arrow is missing, or pointed at the wrong place, an aberration occurs. Identifying gaps in the process is the first step to process improvement.

DSI Core Principles

Now that the ideal structure of DSI has been described, along with typical aberrations to that ideal, it is appropriate to articulate some of the guiding principles that should drive the implementation of DSI at any company. There are three guiding principles to be described:

- DSI should be demand driven.
- DSI should be collaborative.
- DSI should be disciplined.

DSI Should Be Demand Driven

Many years of supply chain research concludes that the most successful and effective supply chains are demand driven. In other words, supply chains are most effective when they begin with the Voice of the Customer. DSI processes should reflect this demand-driven orientation. This principle is operationalized by a focus on the arrow in Figure 7-1 labeled "Demand Forecast." The demand forecast is the Voice of the Customer in the DSI process. However, in too many instances, this customer voice is not as well represented as it should be, because sales and marketing are not as committed to, or engaged in, the process as they need to be. Because of culture, driven by measurement and reward systems, the weak link in many DSI implementations is the engagement from sales and marketing.

DSI Should Be Collaborative

Figure 7-1, the Ideal State of DSI, indicates that inputs to the process come from a variety of sources, both internal and external: sales, marketing, operations, logistics, purchasing, finance, and senior leadership represent the typical internal sources of information, and important customers and key suppliers represent the typical external sources of information. For this information to be made available to the process, a culture of collaboration must be in place. This culture of collaboration is one where each individual who participates in the process is committed to providing useful, accurate information, rather than pursuing individual agendas by withholding or misconstruing information. Establishing such a culture of collaboration is often the most challenging aspect of implementing an effective DSI process. Senior leadership must play an active role in developing and nurturing such a culture.

DSI Should Be Disciplined

When I teach undergraduate students, who often have little experience working in complex organizations, I often make the point that "what goes on in companies is *meetings*. You spend all your time either preparing for meetings, attending meetings, or doing the work that results from meetings." DSI is no different. The core of effective DSI processes is a series of meetings, and for DSI to be effective, the meetings that constitute the core of DSI must be effective also. This means discipline. Discipline comes in several forms. The right people must be in attendance at the meetings, so decisions about balancing demand and supply can be made by people who have the authority to make those decisions. Agendas must be set ahead of time and adhered to during the meetings. Discussion must focus on looking forward in time, rather than dwelling on "why we didn't make our numbers last month." To develop and maintain such discipline, there must be organizational structure in place where someone with adequate organizational "clout" owns the process and where senior leadership works with this process owner to drive process discipline. Again, this points to organizational culture being critically important to DSI process effectiveness.

When these principles are embraced, the "magic" of DSI can be realized. This "magic" is captured in Figure 7-6.

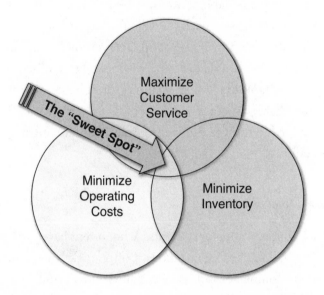

Figure 7-6 DSI "Magic" Comes from Hitting the Sweet Spot

The "sweet spot" shown in Figure 7-6 is the intersection of three conflicting organizational imperatives: maximizing customer service (having goods and services available to customers at the time and place those customers require), minimizing operating costs (efficient manufacturing processes, minimizing transportation costs, minimizing purchasing costs, and so on), and minimizing inventory. This "sweet spot" can indeed be achieved, but it requires DSI implementation that is demand driven, collaborative, and disciplined.

Critical Components of DSI

This chapter is not intended to be a primer on the detailed implementation of DSI. Many other articles and books do an outstanding job of providing that level of detail. However, in my experience of working with dozens of companies, I have observed that five critical components must be in place for DSI to work well:

- Portfolio and product review
- Demand review
- Supply review
- Reconciliation review
- Executive review

Each will now be discussed in terms of high-level objectives rather than tactical, implementation-level details.

Portfolio and Product Review

This step is often absent from DSI processes, but including this step represents best practice. The purpose of the portfolio and product review is to serve as input to the demand review for any changes to the product portfolio. These changes typically come about from two sources: new product introductions and product, or SKU, rationalization. New product forecasting is a worthy subject for an entire book,[2] and as such, I do not dwell on it here other than to say that predicting demand for new products, whether they are new-to-the-world products or simple upgrades to existing products, represents a complex set of forecasting challenges. Too often, new product development (NPD) efforts are inadequately connected to the DSI process, and the result is lack of alignment across the enterprise on the effect that new product introductions will have on the current product portfolio.

Another key element of the portfolio and product review stage of the DSI process is product, or SKU, rationalization. It is the rare company that has a formal, disciplined process in place for ongoing analysis of the product portfolio, and the result of this lack of discipline is the situation described by a senior supply chain executive at one company: "We're great at introducing new products, but terrible at killing old ones." This unwillingness to dispassionately analyze the overall product portfolio on an ongoing basis leads to SKU proliferation, and this leads to often unnecessary, and costly, supply chain complexity. By including product, or SKU, rationalization as an ongoing element of the DSI process, companies have a strong foundation that permits the next step—the demand review—to accurately and effectively assess anticipated future demand across the entire product portfolio.

Demand Review

The demand review is, in essence, the *raison d'etre* for this book. The ultimate objective of the demand review is an unconstrained, consensus forecast of future demand. This meeting should be chaired by the business executive with P&L responsibility for the line of business that constitutes the focus of the DSI process. This could be the company as a whole, or it could be an individual division or SBU of the company. Key, decision-capable representatives from the demand side of the enterprise should attend the demand review meeting, including product or brand marketing, sales, customer service, and key account management. One company that I've worked with has a protocol for its monthly demand review meeting. Sales and marketing personnel are invited and expected to attend and actively participate. Supply chain personnel are invited, but attendance is optional, and if they attend, they are not allowed to participate in the discussion! The intent is that the supply chain representatives are not allowed to chime in with statements like, "Well, we can't supply that level of demand." This protocol is one way that this company keeps the focus on unconstrained demand. In essence, DSI is a consensus forecast of expected future demand.

Supply Review

The supply chain executive with relevant responsibility for the focal line of business being planned by the DSI process should chair the supply review meeting. The purpose of the supply review is to arrive at a capacity forecast, defined as the firm's best guess of its ability to supply products or services in some future time period, given a set of assumptions that are both internal and external. In addition, the supply review is that step in the DSI process where the demand forecast is matched up with the capacity forecast, and any gaps are identified, resolved, or deferred to future meetings.

The capacity forecast is determined by examining a number of pieces of information that are focused on the firm's supply chain. The components include supplier capabilities, actual manufacturing capabilities, and logistics capabilities. Supplier capabilities are typically provided by the purchasing, or procurement, side of the supply chain organization and reflect any known future constraints that could result from raw material or component part availability. Manufacturing capabilities are determined through a number of pieces of information. These include the following:

- Historical manufacturing capacity
- Equipment plans, including new equipment that could increase throughput or scheduled maintenance, and could temporarily reduce capacity
- Anticipated labor constraints, either in terms of available specialty skills, vacation time, training time, and the like
- Improvement plans beyond equipment plans, such as process improvements that could increase throughput

Logistics capabilities can include any anticipated constraints on either inbound or outbound logistics, including possible transportation or warehousing disruptions. Altogether, these three categories of capabilities (supplier, manufacturing, and logistics) determine the overall capacity forecast for the firm.

After this capacity forecast is determined, it can be matched up against the demand forecast produced during the demand review stage of DSI. This is the critical point—where the firm identifies the kinds of gaps described earlier in this chapter. Here is where DSI becomes the mechanism to plan the business. These are the gaps that must be closed: when there is more demand than there is supply, or when there is more supply than there is demand. In some cases, these gaps are fairly straightforward, and solutions are fairly apparent. For example, if there is excess demand for Product A, while simultaneously excess capacity for Product B, it might be possible to simply shift manufacturing capacity from B to A. The answer might be obvious. However, in other cases, as was described earlier in the chapter, the optimal solution to the demand-supply gap might not be so apparent. In those cases, other perspectives must be taken into consideration, which is the reason behind the next step in the DSI process: the reconciliation review.

Reconciliation Review

If a firm were to stop the process with the supply review, as described in the preceding section, it would have a perfectly serviceable, albeit tactical, S&OP process. The reconciliation review, along with the Executive DSI review, transforms S&OP into DSI; here the process is transformed from one that is designed to *plan the supply chain* into one that is designed to *plan the business*. The objective of the reconciliation review is to begin to engage the firm's senior leadership in applying both financial and strategic criteria to the question of how to balance demand with supply. The reconciliation review focuses on the financial implications of demand-supply balancing, and the meeting is thus typically chaired by the Chief Financial Officer responsible for the line of business being planned. Attendees at this meeting include the demand-side executives (sales, marketing, and line-of-business leaders) and supply-side executives (the supply chain executive team), along with the CFO. Its aim is to have senior financial leadership lead the discussion that resolves any issues that emerged from the demand or supply reviews, and to ensure that all agreed-upon business plans are in alignment with overall firm objectives. At this point the discussions that have taken place in previous steps, which have typically focused on demand and supply of physical units, become "cashed-up." In other words, the financial implications of the various scenarios that have been discussed in the demand and supply reviews are now considered. Most unresolved issues can be settled at the reconciliation review, although some highly strategic issues may be deferred to the Executive DSI review, to which we will now turn.

Executive DSI Review

This is the final critical component of the DSI process, and this constitutes the regularly scheduled (usually monthly) gathering of the leadership team of the organization. The chair of this meeting is typically the CEO of the entity being planned, whether that is the entire firm or an identified division or SBU. The overall objectives of the Executive DSI meeting are as follows:

- Review business performance.
- Resolve any outstanding issues that could not be resolved at the reconciliation meeting.
- Ensure alignment of all key business functions.

In other words, in this leadership meeting all key functions of the enterprise, from sales to marketing to supply chain to human resources to finance to senior leadership, come together to affirm the output of all the other pieces of the process, and all functions can agree on the plans that need to be executed for the firm to achieve both its short- and long-term goals. In other words, this is where all functions gather to make sure that everyone is singing out of the same hymnal.

Clearly, for this process to be effective, the right preparation work needs to be completed before each of the scheduled meetings so that decision makers have relevant information available to them to guide their decisions. Also, the right people need to be present at each of these meetings. Both of these requirements lead directly to our next topic.

Characteristics of Successful DSI Implementations

An old quote, attributed to a variety of people from Sophie Tucker to Mae West to Gertrude Stein, says, "I've been rich and I've been poor, and rich is better." In a similar vein, I've seen good DSI implementations, and I've seen bad ones, and good is better! Based on the good and bad that I've seen, here are some characteristics of successful DSI implementations:

- *Implementation is led by the Business Unit executive.* In other words, *DSI cannot be a supply-chain-led initiative if it is to be successful.* The most common reason for the failure of DSI is lack of engagement from the sales and marketing sides of an enterprise. Often, the impetus for DSI implementation comes from supply chain organizations, because they are often the "victims" of poor integration. Many of the 40+ forecasting and DSI audits that have been conducted by our research team have been initiated by supply chain executives, usually because their inventory levels have risen to unsatisfactory levels. The preliminary culprit is often poor forecasting, but poor forecasting is usually just the tip of the iceberg. Poor DSI is usually to blame, and it is often due to lack of engagement from the sales and marketing sides of an organization. So DSI implementations absolutely need commitment of time, energy, and resources from sales and marketing, but the front-line personnel who need to do the work are often unconvinced that it should be part of their responsibility. This problem can be most directly overcome by having the DSI implementation be the responsibility of the overall business unit executive, who has responsibility for P&L, and to whom sales and marketing report.

• *Leadership, both top and middle management, is fully educated on DSI, and they believe in the benefits and commit to the process.* DSI is not just a process consisting of numerous steps and meetings. Rather, it is an organizational culture that values transparency, consensus, and a cross-functional orientation. This organizational culture is shaped and reinforced through the firm's leadership, and for DSI to be successful, all who are engaged in the process must believe that both middle and top management believe in this integrated approach. Such engagement from leadership is best achieved through education on the benefits of a DSI orientation.

• *Accountability for each of the process steps rests with top management. Coordinators are identified and accountable for each step.* Obviously, business unit and senior management have other things to do than manage DSI process implementation, so coordinators need to be identified that have operational control and accountability for each step. DSI champions need to be in place in each business unit, and this should not be a part-time responsibility. Our research has clearly shown that continuous process improvement, as well as operational excellence, requires the presence of a DSI leader who has sufficient organizational "clout" to acquire the human, technical, and cultural resources needed to make DSI work.

- *DSI is acknowledged as the process used to* run the business, *not just* run the supply chain. The best strategy for driving this thought process is to win over the finance organization, as well as the CEO, to the benefits of DSI. Without CEO and CFO engagement, it is easy for DSI to be perceived throughout the organization as "supply chain planning." A huge benefit from establishing DSI as the way the business is run is that it engages sales and marketing. I have observed corporate cultures where DSI is marginalized by sales and marketing as "just supply chain planning, and I don't need to get involved in supply chain planning. That's supply chain's job." However, if DSI is positioned as "the way we run the business," then sales and marketing are much more likely to get fully engaged.

- There is recognition that organizational culture change will need to be addressed in order for a DSI implementation to be successful. Several "levers" of culture change must be pulled in order for DSI to work:

  - *Values.* Everyone involved in the process must embrace the values of transparency, consensus, and cross-functional integration.

  - *Information and systems.* Although IT tools are never the "silver bullet" that can fix cross-functional integration problems, having clean data and common IT platforms can serve as facilitators of culture change.

  - *Business processes.* Having a standardized set of steps that underlie the DSI process is a key to culture change. Without clearly defined business processes, important elements of DSI can be neglected, leading to confusion and lack of integration.

  - *Organizational structure.* Having the right people working in the right organizational structure, with appropriate reporting relationships and accountabilities, is a key facilitator to culture change.

- *Metrics.* People do that for which they are rewarded. What gets measured gets rewarded, and what gets rewarded gets done. These are clearly principles of human behavior that are relevant to driving organizational culture.

- *Competencies.* Having the right people in place to do the work and providing the training needed for them to work effectively are critical elements to successful DSI implementations.

DSI Summary

To bring the discussion full circle, the reader is referred back again to Figure 7-1, Demand/Supply Integration: The Ideal State. Focus on one arrow: the arrow labeled "Demand Forecast." It is my strong belief that without an effective DSI process, an accurate demand forecast is not worth the paper it's written on (or the disk space it takes up on a computer system). But it is also my strong belief that this one arrow represents the critical beginning of the process. World-class DSI processes require world-class demand forecasting, and the companies who are best able to master it will reap the greatest benefits.

Endnotes

1. Mark A. Moon is head of the department of Marketing and Supply Chain Management and an Associate Professor of Marketing in the Haslam College of Business at the University of Tennessee.

2. An excellent book-length primer on new product forecasting is *New Product Forecasting: An Applied Approach*, by K.B. Kahn and M.E. Sharpe, New York: Routledge, 2006.

8

Sell Right, Not More: Leveraging Internal Integration to Mitigate Product Returns

By Diane A. Mollenkopf, Robert Frankel, and Ivan Russo[1]

Demand and supply integration (DSI) refers to the ability of an organization to effectively coordinate the demand functions with the supply functions to co-create maximum value with key customers and the firm itself. Researchers suggest that effective DSI requires the right organizational mindset, knowledge, motivation, and incentive structures to create an integrated firm.[2] Through our own research on returns management over the past decade, we've seen the importance of integrating demand and supply functions within a firm, and even across organizations within a supply chain. In this chapter on internal integration, we specifically address the returns management process to highlight both the need for, and the benefits of, internal integration in firms managing complex activities within and across both forward and reverse supply chains.

The first section of the chapter provides a broad overview of the returns management process. The second section of the chapter provides more specific organizational direction for firms to improve their returns management practices. Throughout the chapter, we rely on the experiences of firms we've worked with as part of our research to highlight the importance of internal integration, as well as the many challenges in creating an integrated organization.

A Returns Management Overview: Inspiring Internal Integration

The activities embodied within the broad domain of supply chain management are often described as a set of processes—a sequence of activities that bring about desired outputs for a firm and its supply chain partners. Cooper, Lambert, and Pagh[3] identified eight processes that must be managed across the supply chain: customer relationship management; customer service management; demand management; order fulfillment; manufacturing flow management; procurement; product development and commercialization; and returns management. Clearly, these processes transcend both demand functions and supply functions within firms that make up a supply chain, and therefore managers must be able to integrate decisions and actions so as to achieve organizational objectives. The discussion in this chapter focuses on internal integration within the returns management process—trying to understand how personnel blend multiple functions into a unified whole.

The Micro Approach: Managerial and Employee Behaviors

In our research, we have taken what is referred to as a *micro approach*. This means we have studied what managers and employees actually *do* in their jobs in managing the returns process, especially as it relates to their interactions within and across organizational functions.[4] Our focus on the process of returns management helps us understand how internal integration occurs and why it's important to firms.

A researcher never knows when a particular interview or discussion will provide a moment of revelation, that is, of significant impact. In our case, a senior director at one firm coined a phrase that has stuck with us over the years and forms the guiding principle of our discussion in this chapter. "Sell Right, Not More" embodies the philosophy of how to create a more internally integrated firm. In his firm—an international consumer appliance manufacturer that we call Action Appliance (pseudonym)—the marketing/sales team had traditionally been incented to "sell more" to retail customers. From a revenue-generating perspective, that made a lot of sense. The manufacturing and distribution operations group, however, had a different reaction to the "sell more" credo. The operations personnel saw the end-of-season returns from retailers. They had to manage the reverse flow of product that came back from customers, which increased transportation, warehousing, and inventory costs. Sometimes the returned product could be held over for the following season, but when new models were planned for introduction, the now obsolete inventory had to be discarded—either sold at deep discounts or broken down for parts and recycling. The "sell more" approach to working with customers was increasing the firm's cost of goods sold and logistics costs, negatively impacting the firm's profitability. But the firm didn't see the financial implications of "selling more" when the demand side of the organization and the supply side of the organization were being held to different goals: one related to revenue enhancement, the other to cost minimization. In other words, the demand function and the supply function were not integrated.

When the new Logistics Director came on the scene and began analyzing the costs that were created through the returns management process, the company started to understand that "selling right" might be more important than "selling more." The total costs of managing return products from retail customers were significantly impacting the company's bottom line. In addition to the demand and supply challenge, the question arose as to how to convey the benefits of "selling right" to the firm's powerful wholesale and retail customers. "Selling right" meant that initially, sales personnel were accompanied by operations and finance personnel when meeting with customers so that pricing terms and order quantity discussions were based on profitability impact, not just revenue impact. Delivering direct to a retailer's store from overseas created significant cost differences compared to delivering to the retailer's distribution center from Action Appliance's central U.S. warehouse. When Action Appliance fulfilled the retailer's order from its central warehouse, they had greater visibility of the inventory in the retailer's network and could help manage that inventory. Inventory visibility was compromised when retailers chose to have direct store delivery from the overseas manufacturing location. Thus, Action Appliance built different pricing structures and related return policies into purchase orders with retailers, depending on the chosen delivery method.

This new approach wasn't just about the firm benefitting; by taking the "Sell Right, Not More" approach, retail customers received improved service and better prices, so their own performance also improved. Action Appliance went from being "just another supplier" to major retailers, and became a key supplier to many of its retail customers because it could demonstrate the value it was bringing to customer relationships when it took an integrated approach to managing those relationships.

Ultimately, the company was restructuring its business processes so that demand and supply functions were jointly focused on making profitable sales. By creating a common profitability goal across the company, marketing and operations teams went from working at cross-purposes to each other and began to understand how to jointly contribute to the organization's overall success. A new organizational mindset, coupled with a willingness to share information, enabled different kinds of decisions to be made about how to sell to customers and how to fulfill customer orders. In this particular case, as it relates to returns management, the reduction of return volume was the driving force to enhanced internal integration. The key to understanding and improving the returns management process is not whether returns occur (they will), but how a company views the role of returns in its overall business strategy.

The Returns Management Process

The importance of a process perspective to managing supply-chain-related activities cannot be underestimated. While traditional management approaches have focused on functional expertise, the resulting silos have often been at the root of the divide between the demand and supply sides of an organization. In the example of Action Appliance, the "sell more" approach became so embodied in the firm's approach to dealing with customers that bridging those silos was a major challenge. Only when the Logistics Director coached the firm's managers on the wide range of internal activities that took place in supporting their customers' needs was the firm able to transcend the silos and develop a process perspective to managing returns that ultimately benefitted both the firm and its customers.

The returns management process's span of decision making suggests that those involved in its activities must share a common understanding of the process, its goals and objectives, and be able to communicate across functional boundaries to ensure that returns are managed as efficiently and effectively as possible.[5] This process also includes gatekeeping (screening of a return request and/or returned product) and avoidance (developing and selling a product to minimize return requests). Managing activities such as reverse logistics, refurbishment, and remanufacturing are increasingly being performed by a third-party firm, which extends and complicates the span of control to the interfirm realm. Consider the various functions involved in the returns management process:

- *Marketing/Sales.* Marketing and sales managers often determine the return policy when negotiating the terms of trade with a customer. Subsequently, salespeople or their managers often authorize specific product returns when initiated by customers. Salespeople also negotiate new product placements with retailers, which often involve taking back older stock in order to place newer stock on store shelves. Sales personnel are also tasked with selling returned products, sometimes into secondary channels.

- *Legal.* The legal team gets involved at the contract stage of negotiating terms of trade with new customers (or revising contracts with existing customers). Often, issues of gatekeeping are addressed at the contract stage of negotiating with customers. This occurs when firms recognize that some products are more costly to return and process than they are worth. Therefore, customers may be encouraged to keep the product (while still receiving credit) or be instructed to destroy-in-field. Gatekeeping is particularly important for promotional inventories that may have little value after a promotion period.

- *Customer Service.* Customer Service Representatives, as the frontline interface between customers and the firm, often receive the initial request for return from customers. Depending on the size, nature, and financial impact of the return shipment, the representative will initiate an authorization process before product is shipped back from a customer, often in conjunction with Marketing/Sales and Finance.

- *Logistics.* Transportation personnel may have been involved in actually arranging the shipment, but sometimes the customer arranges for physical movement of the product. Warehouse personnel receive the returned product when it arrives at the back dock. Paperwork (increasingly electronic) must be matched with the customer initiating the shipment, and each product inspected and verified, comparing what was received with what the authorization form stipulated in terms of quantity and quality of product being returned. Inventory records must then be updated to record the receipt of the returned product. Product that can be returned immediately to stock for sale is returned to storage. Other product is moved to refurbishing, remanufacturing, or disposition areas for further processing.

- *Refurbishment/Remanufacturing.* Product that cannot immediately be returned to inventory will be inspected for refurbishing or remanufacturing potential. Products that cannot be brought back to a "like-new" status can be broken down for parts recapture. Parts may be selected to support the after-sales service teams, or be designated for component recycling or disposition.

- *Accounting.* The credit team is charged with reconciling the product receipt records from the warehouse personnel against the initial return authorization before issuing credit to the customer.

Integrated Forward and Reverse Flows

Returns have long been treated as the "ugly stepchild" of business activity, or perhaps just a necessary cost of doing business. Indeed, most business activity is focused on selling product *to* customers and moving product *to* customers. It's no wonder that returns receive so little attention. Yet, when considering issues of demand and supply integration, both forward and reverse flows of product must be built in to a company's strategy. From a purely financial perspective, the numbers are rather staggering and clearly indicate the need for management attention. According to the Reverse Logistics Association, in the U.S. retail sector alone, the volume of returns annually is estimated at 6 percent of $3.5 trillion total annual retail sales. Supply chain costs associated with reverse logistics have been estimated to average 7–10 percent of cost of goods.[6] Moreover, when only considered as a cost of doing business, returns management gets managed as a very operational activity, focused solely on cost reduction at the warehouse, distribution center, or retail level. In such cases, firms are missing out on strategic opportunities to create value for themselves and with their customers.

To move beyond the narrow focus of considering returns management as just a cost of doing business, some historical perspective may be beneficial. In the first half of the twentieth century, early scholars of marketing developed the concept of "marketing flows" that illustrated activities both within a firm and across a marketing channel.[7] The "flows" concept set the stage for contemporary conceptualizations of managing processes across functional boundaries. Particularly salient for our purposes are the elaboration of physical flows, information flows, and financial flows that occur within a firm as well as across channel (supply chain) members.

- *Physical flow.* The actual movement of goods, from raw materials, parts, and components to finished goods that move ever closer to the end customer.
- *Information flow.* The exchange of relevant information about the goods flow as it moves across the supply chain. This might include information such as product specifications, order size, shipping date, delivery date/location, pricing, and sales terms.
- *Financial flow.* The exchange of money between channel members as goods move and ownership is transferred from one company to another.

The predominant conceptualization of the supply chain is represented as a forward flow of products moving to customers and consumers, as depicted in Figure 8-1. In this simplified conceptualization of a supply chain, the product flow, information flow, and financial flow are relatively easy to visualize. Within the integrated firm[8] in the supply chain, the supply functions of procurement, operations, and distribution (logistics) must coordinate and cooperate with the demand functions of marketing and sales in order to meet the needs of the customers and, ultimately, the end consumers. Information flows bidirectionally, both across firms and within firms to facilitate the efficient and effective flow of product across the supply chain as it moves to the customer. The financial flow is represented as moving from customer to supplier, as funds are exchanged for goods as products flow toward the end consumer. While funds are exchanged across organizations, the accounting function within a firm must liaise with the demand and supply functions to invoice customers in a timely and accurate manner, as well as to pay suppliers for received goods.

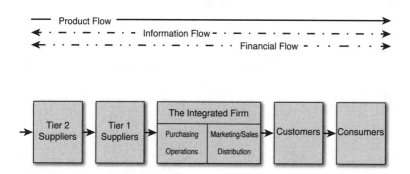

Figure 8-1 The Supply Chain as a Forward Flow of Product

Firms are increasingly recognizing the need for greater emphasis and strategic thinking regarding the role of the reverse supply chain as part of an overall supply chain strategy. This is occurring for many interrelated reasons. First, firms are taking advantage of the inherent value of components and/or parts that can be reused in the manufacturing process to reduce the cost of inputs. In industrial markets, engines and machinery parts have been recovered and refurbished or remanufactured for decades because of their intrinsic value and long life cycles. In the computer industry, manufacturers have been recovering used products to recapture the precious metals contained within each unit. In larger, industrial computer equipment, recovered and/or refurbished parts may be merged into field support teams as repair components. Value recapture may also involve closed-loop systems of packaging materials, such as barrels, storage containers, totes, or racks that can be recovered and reused multiple times for significant cost savings. These examples illustrate how the cost of goods sold can be reduced—sometimes dramatically—to positively impact a firm's profitability. Value recapture may also involve revenue rather than cost reduction. In retailing, firms realize that processing returned products efficiently means potential sales to customers during the current season (an unworn garment purchased online, for example), which may have gone unrealized if returns were poorly managed. Efficient processing may also enable geographic or store-level repositioning of returned inventory to balance supply and demand. Thus, firms are able to increase profitability through reduced cost of goods sold and/or increased revenue by recovering products, parts, and even packaging.

Environmental concerns present a second reason to develop reverse supply chains. Firms are actively reducing their environmental impact by keeping products out of landfills through product recovery programs. Recaptured products or equipment can be reused or recycled to reduce waste and reduce demand for virgin resources. On the resource front, growing concern for scarce resources further augments the rationale for developing reverse supply chain processes. As the global population increases, and the burgeoning middle classes demand increasing amounts of consumer goods, global concern is rising about the long-term availability of some natural resources. For example, certain nonrenewable resources, such as metals and minerals, are increasingly difficult to obtain. Many firms are finding that recapturing product from the marketplace at end-of-life or end-of-use provides a ready (and relatively inexpensive) source of valuable inputs. Consumer electronics manufacturers, for example, not only recognize the value potential of recapturing the precious metals contained within each unit, they also recognize the importance of mitigating supply chain disruption risk due to scarcity of such inputs.

Regulation and political action are also closely related to the environmental initiatives just mentioned. In the European Union, for example, extended producer responsibility legislation known as the Waste Electronic and Electric Equipment (WEEE) Directive[9] requires any manufacturer or distributor of electronic/electric equipment to ensure that its products do not end up in landfills. On the scarcity front, exports of rare earth metals are limited by foreign governmental policies, making reverse supply chains increasingly attractive to firms reliant on these inputs. For example, to mitigate scarcity-related risk in its supply chain, Honda Motor Company has implemented efforts to recover and recapture rare earth metals in the batteries of their hybrid automobiles[10] by creating a closed-loop supply chain to manage the reverse flow of products. In the United States, the recent passage of the Dodd–Frank Bill requires companies that manufacture any products containing tin, tantalum, tungsten, or gold—referred to as 3TG, or "conflict minerals"—to document the source of these minerals because of their potential origin from Democratic Republic of the Congo (DRC), where human rights violations are rampant.[11] The regulation has the effect of creating scarcity of documented, conflict-free 3TG. Firms may find it easier and less risky to recapture products from the marketplace and reuse these minerals than risk unknown sources that may emanate from DRC.

Whether the driving force for developing a reverse supply chain is value recapture, or stems from an external source related to environmental impact, scarcity, or regulation, or some combination of all these forces, the implementation of a reverse supply chain coupled with a firm's traditional forward supply chain requires a significant level of internal integration. Forecasting efforts need to address both the demand for products as well as the return of used/old product. Managing the timing, traceability, and quality of materials flow is critical for production efforts, especially when virgin materials and recovered materials must be merged into the production process. Blended inventories must also be carefully managed to support sales efforts across a variety of primary and secondary markets.

But the reverse supply chain is a tough nut to crack; it is much more complicated to manage than traditional forward supply chains. Recovering products from the marketplace can be very challenging, particularly in consumer markets where manufacturers may not know who purchased their products. Products at end-of-use or end-of-life are much more costly to manage in the supply chain as well; the uncertainty of their valuation complicates the cost equation. Additionally, transportation, inventory, and storage considerations can be problematic. Although containerization, intermodal innovations, and connectivity have helped firms enhance the efficiency of the forward flow of products across global supply chains over the past several decades, moving products in reverse remains challenging. Products are no longer uniform, in protective packaging; opportunities to gain economies of scale are less frequent; and products may or may not travel the same route backward as they did on the forward journey to customers.

This complexity is particularly salient as omnichannel retailing evolves to include more consumer touch points. As consumers have an increasing number of means to engage with a company, they also have more opportunities to return products in a variety of ways. But consumer returns are only one form of returns that must be managed. Figure 8-2 highlights the complexity of the reverse channel, while summarizing the many types of returns, which can reenter the reverse supply chain from multiple points within the chain. In fact, many firms are establishing external processing facilities that route products away from the forward channel to a processing center to achieve valuable economies of scale as early as possible in the reclamation and disposition stages of products' reverse journey. Ultimately, products returned through the reverse supply chain will be reinserted into the forward supply chain in some form (as resalable new products; as refurbished or remanufactured products; or in parts form).

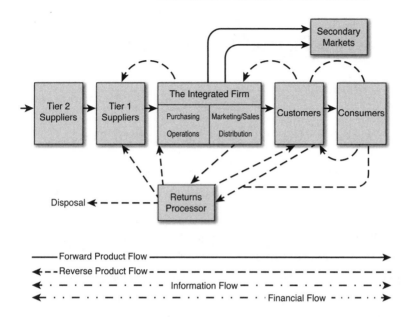

Figure 8-2 The Integrated Forward/Reverse Supply Chain

TYPES of RETURNS[12]

Consumer:	Due to buyer's remorse or product defects.
Marketing:	Business-to-business returns, such as unsold product from retailers.
Damage:	Damaged in transit, warehouse, or retail store; initiated by a channel member, not the end consumer.
Asset:	Recapture and repositioning of an asset such as a reusable container or leased equipment.
Material reclamation:	Components recovered for reuse.
Product recalls:	Voluntary or government-mandated returns due to safety or quality concerns.
Environmental:	Disposal of hazardous materials or abiding by environmental regulations.

Only by understanding returns management as part of a larger process of managing forward and reverse flows of product, information, and money across supply chain organizations can managers appreciate the need for internal integration. In the next section we turn our attention to discussing how managers improve the internal integration within their firms when managing both forward and reverse product flows.

The Challenge of Managing Returns

As noted earlier in the chapter, integration refers to the organizational initiative to blend departments/functions into a unified whole. While the conceptual boundary between two functions is evident, bridging that boundary is far more challenging and can be contentious in practice. When we began our research a number of years ago, our initial focus was on the marketing/logistics interface and how people in these functional areas interacted when managing return products. Very quickly, we realized the need to include other functions that played a role in the returns management process. In other words, the scope of the challenge we were investigating was much broader than it initially seemed.

Up to this point in the chapter, we have treated internal integration as a singular concept that stands on its own, having provided no elaboration as to its composition. In actuality, integration is a multidimensional concept whose components are highly interrelated:[13] (1) alignment, (2) cross-functional integration, and (3) intrafunctional integration. Of these three dimensions, alignment and cross-functional integration are most salient for our discussion in this section; additionally, they represent a critical interface with the three flows of product, information, and finances that will be explored as well. As our research has evolved over the years, we now recognize that internal integration requires *both* cross-functional integration as well as alignment, and that each is a necessary but insufficient condition when considered alone.

Cross-functional integration can be paraphrased as a process of interdepartmental interaction and interdepartmental collaboration that brings departments together into a cohesive organization.[14] Cross-functional integration, because it is focused on functions that should work in tandem, is conceived of as a *horizontal* domain of internal integration. Although interest in cross-functional integration can be traced back at least 35 years in the academic literature, a concentrated focus on cross-functional integration by scholars and practicing managers has become much more common in the past 15–20 years.

Alignment is defined as the consistency of objectives across functions, as related to the firm's larger corporate strategic goals.[15] Thus, alignment is concerned with a broader organizational scope; in other words, activities and decisions not only occur across functions, they occur across levels within the firm—from the dock worker receiving and processing the return, to the senior vice presidents who set policies and strategies. Alignment, because it is focused on the hierarchical levels within a firm, is conceived of as a *vertical* domain of internal integration. Perhaps the best way to understand the importance of alignment is the saying from the famous consultant and writer Peter Drucker, who noted that "making good decisions is a crucial skill at every level of the organization."[16]

Cross-Functional Integration

We start by examining the horizontal domain of internal integration. A key strategic decision for firms is to determine whether to avoid or to encourage returns. Although this will be somewhat dependent on the nature of the product (laptops have much more value recapture potential than lipsticks, for example), the decision will have implications for supply chain design and the nature of the interaction across functional units in an organization. Take a book publishing company we worked with in Italy, for example. The book publishing industry is fraught with demand uncertainty. After an author is contracted to write a book, the publisher needs to determine how many units to print in the initial production run. Too few for a potential best-selling book, and sales will be sacrificed. Too many, however, undermines profitability by committing resources to production and distribution that will never be recouped through future sales. The market structure in Italy is transforming from predominantly mom-and-pop independent retailers to larger chains, but at the time of our research, the mom-and-pop retailers still represented a significant portion of the market.

Traditional publishing approaches were to forecast demand for a book, order the print run, and distribute to retailers. If a book sold faster than anticipated, a second print run would be ordered. The printing factory ran on a locked-in production schedule, and urgent requests for reprinting caused significant disruptions. If a book sold less than expected, however, the retailers suffered with unsold stock on their shelves. The publisher wanted to sell more books to the retailers, and although the retailers certainly wanted faster-moving books to entice a steady stream of customers into their stores, their often limited capital meant they had no funds to purchase more books unless the unsold books could be returned to the publisher for credit. Thus began a costly cycle of publishing books that would only be returned to the publisher in order to sell more books to retailers. The returned books sat in warehouses for three years and were then destroyed.

The publisher recognized disconnects across its demand and supply functions and the resulting profitability impacts of its existing system. Adopting a "Sell Right, Not More" approach required a major overhaul of the production process to create a more flexible, demand-pull approach to publishing. When a new author was signed, an initial small run of books would be published and sent to retailers. If demand continued, the factory could more easily match supply with demand through its new flexible production process. This required significant cross-functional integration across marketing, production, distribution, and financial roles in the organization to agree on how to create a "Sell Right, Not More" system. Ongoing operations required continual information sharing across functions to time the production to the needs of the market. Overall, by focusing on how to avoid returns, the company was able to reinvent itself and boost profits significantly, while also providing greatly enhanced service to its retailer base.

While the book publisher sought to avoid returns, another company we have worked with sought to encourage returns. A large Italian automotive parts distributor (a subsidiary of a global automotive manufacturer) encouraged returns in order to provide better service to its dealers in an increasingly competitive marketplace. Due to increasingly stringent legislation in Europe regarding sustainable disposal of car parts, the distributor developed a system to make it easier for dealers to return parts to it. This marketing strategy also required a supply chain capability that ensured a simple and effective mechanism to replace parts so dealers could serve their own customers better. The system was designed as a closed loop, and required cross-functional integration across sales, production, logistics, and accounting to properly balance and replenish parts inventory at multiple forward and reverse locations in the company's supply chain. Rather than simply having production generate more inventory "just in case" a sale occurred, the system encouraged dealers to return unneeded inventory in a continuous, rapid manner to place and balance parts where and when needed. For this company, the principle of selling the "right" inventory and not merely "more" inventory at the distribution and dealer (retail) level has been the key to successful supply chain management. This is another example of cross-functional integration across the demand and supply divide within an organization that ultimately has changed the way the company competes in its marketplace.

Alignment

We continue our discussion with consideration of the vertical domain of internal integration. Two aspects are paramount:

understanding the total costs of returns and ownership of the returns management process.

Transitioning to a "Sell Right, Not More" philosophy requires a solid understanding of the total costs of managing returns. In our prior discussion of the Italian book publisher, senior management realized that the tremendous costs of production coupled with the costs of lengthy storage of unsold books before destroying them far outweighed the marketing benefits of creating and maintaining sufficient inventory to ensure retailers would not stock out. This total cost analysis enabled both the demand and supply sides of the organization to integrate their efforts to better serve customers while also improving the organization's profitability. In too many situations, we've discussed the cost of returns with managers, only to realize that upper management has no idea of the true cost of returns. Moreover, the visibility of such costs and their associated impact is not shared throughout all organizational levels. In one company that had recently undertaken a project to better capture the true cost of returns, the CFO bemoaned to us that he now understood that "returns cost me a lot of money." The financial impact of returns management clearly validates its importance in the supply chain.

A particularly impactful example of a failure to understand the total cost of returns took place at a large office supply manufacturer in the United States. In this firm, the vice president of sales was concerned only with revenue of his sales force and was not held accountable for any inventory implications of his sales teams' decisions regarding accepting return product back from customers. Likewise, on the operations side of this firm, returns costs were not separate line items in the company's warehouse P&L statements. Return transportation costs were just part of the overall transportation bill. Warehousing labor was not tracked by departmental area, so no one in management knew how many labor hours were being spent processing return goods that were then stored for an indeterminate time because the returned products were no longer salable in the marketplace. The inventory costs and related warehouse costs were also buried in broad cost categories. It's no wonder that there was very little understanding of total cost at this firm, with regard to either the forward or the reverse product movements. Only the plant-level accountant we spoke to could see the overall problems. She boldly stated that for the company to make more profit, the solution was to "sell less." She said this because she saw the cost of all the returns that came back due to product proliferation issues, and the "sell more" approach taken in isolation by the sales teams. Sadly, no one in upper management (especially on the sales side) was interested in listening to her argument. Her case is not unique; in fact, it is quite common. Not only must total cost be understood, but it must be communicated across all levels of the organization. We have experienced numerous similar situations in other firms, across multiple industries and geographies throughout our research over the years.

Happily, we have also conducted research with firms that do exhibit good, consistent internal alignment that is evidenced by a solid understanding of the total cost of managing returns. In Europe, we have worked with a distributor subsidiary of a global pharmaceutical firm engaged in research and development, marketing, manufacture, and distribution of pharmaceutical and health-care products. Clearly, this firm's recognition for strategic thinking with regard to the reverse supply chain is driven by regulation; it operates within one of the most highly regulated industries in the world. Because this company is listed on the New York Stock Exchange, the European subsidiary is subject to Sarbanes–Oxley (SOX) regulations. Although SOX is fundamentally about accountability, higher levels of internal integration are required to ensure conformity with the legal requirements. The company also operates with extremely long cash-to-cash cycles, which exposes the firm to significant financial risk if forward and reverse product flows are not managed well. Product flows (based on "use by" dating) and information flows (for example, disposition instructions) are thoroughly and consistently communicated via explicit policy and procedure up and down the organization. Notably, understanding the cost of a return in this firm is not just financial in nature. The implications of poorly managed returns are also recognized and understood in terms of damaged consumer loyalty and brand perception.

The second aspect of alignment relates to the issue of ownership. Just as functional areas have leaders, organizations need to assign leaders to each process to ensure that everyone's activities up and down the organization are consistently supporting the goal of the process. In our returns management research, when we asked managers, "Who 'owns' the returns process at your firm?" we were usually met with blank stares. At one firm, a director bluntly stated "Nobody. To be honest, nobody and everybody." When nobody is charged with managing the returns process (or any of the supply chain processes that span functional departments within a firm), companies are compromising their ability to integrate demand and supply and their financial performance.

One firm we've worked with is a U.S.-based floor coverings manufacturer that operates globally. The firm exhibits a fanatically devoted commitment to ownership of the returns management process and remanufacturing process via a closed-loop business model based upon returns of flooring products (for example, its remanufacturing technology is based upon used flooring as the raw material for certain production processes). Clear responsibility for managing these two interrelated processes is understood across the global organization; clear responsibilities have also been infused across all levels of the organization, from the salesperson who builds in product recovery to the sales of new flooring, to the logistics operator charged with pickup and transfer of the used flooring back to the remanufacturing plants, to the senior managers who are held responsible for the service and cost elements of the entire closed-loop process. The approach is part of its competitive positioning to not only attract new customers but to develop increased loyalty among existing customers. It can be successful only because while clear responsibility lines are established, an element of process ownership has also been diffused throughout the organization. In contrast to the office supply company in which "nobody and everybody" owned the returns process, this company provides a positive example of what can be done when "somebody and everybody" owns the process.

It is valuable, of course, to assign the ownership of a supply chain process to an individual manager or director and his or her team. But equally important is having the means in place to ensure accountability for that process. Accountability strengthens ownership. Such responsibility requires that proper metrics be in place throughout the organization at all levels; such metrics (which ones, how measured and how frequently, and appropriateness) must be jointly agreed upon by all functions that "touch" the returns management process. In our research, when we asked directors, managers, and front-line employees, "How are you measured on returns management?" we were continually amazed by the absence of metrics—or if metrics existed, the lack of their usage to construct policy and procedures. In contrast, we saw accountability exhibited in a most impressive manner in the global floor covering firm. Its accounting procedures utilize a triple-bottom line approach, wherein the metrics include profit, environmental, and social impact. Clear metrics that are aligned across the organization, and "translated" to specific functional and cross-functional activities, support the previously mentioned infusion of responsibility throughout the organization and create a means for accountability. This firm evidences a remarkable consistency of mission, ownership, and accountability.

With regard to policy and procedure, for example, when a large return shipment is received from a customer that requires significant labor hours to process, will the contractually imposed 24-hour processing window be able to be met? Does it need to be met in every circumstance? What kind of advance warning did the warehouse crew have to prepare for the necessary labor? Can initial credit be approved within the approved time window, but allow time later for further processing? Only when someone owns the process and is accountable for its results can the necessary information and subsequent decisions be made that cross the hierarchical decision-making levels in a firm in order to provide an aligned solution that supports corporate goals.

In any firm, an important corporate goal is satisfying its customers. In fact, a typical response we've met with from sales directors is that strict return policies compromise relationships with customers. However, as the managers at Action Appliance realized, customer relationships can actually be enhanced when return policies become a more integrated component of the relationship. Only when the new Logistics Director took ownership of the returns process could typical cost/services trade-offs be reconciled and their rationale communicated up and down the organization.

Synchronizing the Flows

The vertical and horizontal components of cross-functional integration and alignment that have been discussed must be synchronized with the previously described product, information, and financial flows. Such synchronization, which implies a significant level of visibility across firm functions, is especially challenging in the technology-driven world in which businesses currently operate. We'll consider why this is true.

In the pre-Internet age, return products arrived at a company's back dock with paperwork attached. After processing, the company issued appropriate credit to the customers. The Internet age and its associated technologies have changed the nature of these flows dramatically, but have not in any way diminished the fact that "Sell Right, Not More" is a viable strategy to achieve internal integration. We'll explore those changes in flows in more detail.

On the positive side, information can now flow much faster than the product itself. After a return shipment has been authorized, information about the return in terms of units, reason codes, prices, routing, carriers, and so on can all be available to the accountants and the warehouse staff who will be processing the shipment. This means that staffing levels can be better planned because advance notice can be provided about the return shipment, including the ship date and expected receipt date. On the negative side, the potential to provide faster information flow is of little benefit if the procedure of proper (that is, timely, accurate) transmittal is not executed. "Christmas every day" was a phrase we've heard many times over the years as warehouse managers describe the nature of the returns work in their facilities. Every day the trucks pull in with surprises—return shipments arrive that have not been anticipated because no one remembered to enter the return authorization into the company's database or thought that it wasn't really important to warehouse or distribution center activity. In today's technology-connected world, this should be an "easy fix" in a company applying cross-functional approaches. Although surprises will still occur when a customer ships back product without having first secured a returns authorization, this is a relationship management problem that needs to be addressed, rather than being an information flow problem.

While information technology can and should help firms better manage the information flows and coordinate the information flow with the product flow, the financial flow may become increasingly problematic for firms that do not have a well-integrated returns management process. In many firms we've visited, key customers have become so frustrated with the companies' slow processing of return credits that they have resorted to "taking" the credit by simply short-paying an existing invoice. This is particularly problematic when funds are transferred electronically, creating reconciliation problems with the accounting departments that are responsible for issuing credits to customers. Moreover, there is a high likelihood that the legal function will be drawn in to resolve these incidents, which only further complicates managing the customer relationship. In a number of the firms we have researched, accounting personnel have made it standard practice to not only routinely visit the warehouse/distribution center floor to resolve credit claims, but to actually check in and process a large return in anticipation of reconciliation problems. When customers start taking a credit they think they are due, it becomes very difficult after the fact to adjust the credit if there is any kind of dispute or discrepancy involved—perhaps most important, after-the-fact resolution strains a firm's ability to ensure that a well-designed, consistent policy "face" is presented to customers.

Thus, internal integration involves understanding and managing both cross-functional integration and alignment—as well as synchronizing them with product, information, and financial flows. Perhaps most important, in combination, they provide a basis for managing a firm's external customer relationships.

Thoughts and Observations

The discussion on internal integration as focused on the returns management process within a firm highlights several managerial actions that can be taken to enhance a firm's internal integration.

- *"Sell Right, Not More"* provides a guiding mantra for reconceptualizing how the demand side and the supply side of an organization need to work toward a common goal to enhance firm performance.

- *Integrate process with firm strategy.* The key to understanding and improving a process such as returns management is how a company views that process within its overall business strategy. Although reverse flows have traditionally been overlooked, significant opportunity exists to enhance value creation for customers, suppliers, and the firm itself through better integrated and synchronized forward and reverse supply chain flows.

- *Focus on the total cost of returns.* Working toward a common goal requires that both revenue and cost implications of returns management are understood. Although such an understanding is primarily directed toward enhancing internal integration, managers may also find that understanding the total cost of returns enables better customer relationship management.

• *Own the process; be accountable.* Just as functional departments have, processes must also have leaders responsible for process performance and teams that are held accountable to procedures and expected outcomes.

Internal integration is an easy concept to understand, but a challenging concept to implement within a firm. The discussion in this chapter has used the returns management process to explore the importance and impact of internal integration. The vertical and horizontal dimensions of internal integration, with an eye toward synchronizing managerial decisions and activities related to product, information, and financial flows, have been a productive lens for us during the past decade. Our research experiences with firms around the globe have provided multiple examples of firms that have figured out how to "Sell Right, Not More," as well as multiple examples of firms that are struggling with the transformative journey toward improved internal integration. The discussion in this chapter provides managers with a useful map on the journey to "Sell Right, Not More."

Endnotes

1. Diane A. Mollenkopf is the McCormick Associate Professor of Supply Chain Management in the University of Tennessee's Haslam College of Business. Robert Frankel is the Kip Professor of Marketing and Logistics at the University of North Florida, and Ivan Russo is an Associate Professor of Marketing and Supply Chain Management at the University of Verona, Italy.

2. Tate, W.L., D.A. Mollenkopf, T.P. Stank, and A. Lago da Silva. 2015. Integrating Supply and Demand. *MIT Sloan Management Review* 56(4): 16–18.

3. Cooper, M.C., D.M. Lambert, and J.D. Pagh. 1997. Supply Chain Management: More Than a New Name for Logistics. *International Journal of Logistics Management* 8(1): 1–14. Rogers, D.S., D.M. Lambert, K.L. Croxton, and S.J.

García-Dastugue. 2002. The Returns Management Process. *International Journal of Logistics Management* 13(2): 1–18.

4. Oliva, R. and N. Watson. 2011. Cross-Functional Alignment in Supply Chain Planning: A Case Study of Sales and Operations Planning. *Journal of Operations Management* 29(5): 434–448.

5. Rogers et al. The Returns Management Process. 1–18.

6. Rogers, D.S., R.S. Lembke, and J. Bernardino. 2013. Reverse Logistics: A New Core Competency. *Supply Chain Management Review,* May–June, 40–47.

7. For a brief overview, see Bowersox, D.J. and E.A. Morash. 1989. The Integration of Marketing Flows in Channels of Distribution. *European Journal of Marketing* 23(2): 58–67.

8. This term has been adapted from the Integrated Supply Chain Framework found in Bowersox, D.J., D.J. Closs, M.B. Cooper, and J.C. Bowersox. 2013. *Supply Chain Logistics Management*, 4th edition. New York, NY: McGraw-Hill Irwin.

9. Anonymous. 2012. Directive 2012/19/EU of the European Parliament and of the Council of 4 July 2012 on Waste Electrical and Electronic Equipment (WEEE). *Official Journal of the European Union* 55(24 July): 38–71.

10. Honda Motor Co. 2012. Honda to Reuse Rare Earth Metals Contained in Used Parts. News Releases 2012. Tokyo, Japan, April 17, available at www.world. honda.com/news/2012/c120417Reuse-Rare-Earth-Metals/index.html.

11. Coulter, C. and N. Burton. 2014. Conflict Minerals and Corporate Supply Chains: The Challenge of Complying with Dodd-Frank. *Supply Chain Quarterly.* Accessed online at http://www.supplychainquarterly.com/topics/ Procurement/20140304-conflict-minerals-and-corporate-supply-chains-the-challenge-of-complying-with-dodd-frank/.

12. Lambert, D.M., editor. 2014. *Supply Chain Management—Processes, Partnerships, Performance*, 4th edition. Ponte Vedra Beach, FL: Supply Chain Management Institute.

13. Pagell, M. 2004. Understanding the Factors That Enable and Inhibit the Integration of Operations, Purchasing and Logistics. *Journal of Operations Management* 22(5): 259–487.

14. Kahn, K.B. and J.T. Mentzer. 1996. Logistics and Interdepartmental Integration. *International Journal of Physical Distribution & Logistics Management* 26(8): 6–14.

15. Pagell. Understanding the Factors That Enable and Inhibit the Integration of Operations, Purchasing and Logistics. 259–487.

16. Drucker, P.F. 2006. *Classic Drucker*. Cambridge, MA: Harvard Business School Publishing Corporation. p. 120.

9

Supplier Integration via Vested Relationships

By Kate Vitasek[1]

Today most experts agree that typical organizations spend between 40 and 80 percent of revenue with suppliers that help them develop, manufacture, sell, and service their goods/services. For example, the automobile industry spends 70 percent of its revenue with suppliers.[2] And if you are like most, this means roughly half of your procurement spend is on services that require a more sophisticated approach to sourcing.

For the most part, procurement professionals use transaction-based approaches as the foundation for buying goods and services. However, the nature of business is shifting. First, the C-suite executives around the world took the advice of legendary management guru Peter Drucker to, in effect, "Sell the Mailroom,"[3] which started a subtle shift where organizations began to outsource anything that was not a core competency. As Thomas Friedman has famously pointed out, the world is now indeed "flat," and outsourcing has taken hold as a formidable way for organizations to focus on what they do best.[4]

The rise of outsourcing has been met with organizations seeking new ways to work with suppliers to help drive innovation that creates a competitive advantage. Dr. Robert Handfield and Gerard Chick, authors of *The Procurement Value Proposition*, point to supplier collaboration as "the new way." They state: "The old adversarial posture of procurement is as outmoded as it is inappropriate."[5] Handfield and Chick call for a clear and definitive break between older or past-generation procurement practices and those of today. In short, they argue that everything that has been done and learned in the past will not be useful in the dawn of procurement's new value proposition.

Handfield and Chick are not alone in their conclusion.

Academic research on collaboration has exploded over the past 20 years. Research at the University of Tennessee shows that innovation and collaboration are not mutually exclusive; they feed and build on each other. Innovation happens *through* collaboration. And the best organizations not only *say* they want innovation and collaboration; they go all-in and *contract* for it.[6]

So what does this mean for today's procurement professional? The tried-and-true tools and tactics adopted over the past 30 years are no longer as effective as they once were. If firms are going to compete "supply chain to supply chain," shouldn't all the links in the supply chain work together? But when and how does a procurement professional know when to invest time and energy building highly collaborative relationships? The answer lies in understanding where your sourcing initiative falls along the sourcing continuum and in architecting the appropriate sourcing business model for your supply chain.

The Sourcing Continuum

For centuries, organizations have considered procurement as a "make versus buy" decision. This is especially true as organizations began to explore outsourcing. Many assume if they "buy," they should use competitive "market" forces to ensure they are getting the best deal. In doing so, the default approach is to use a transaction-based model. This works well for simple transactions with abundant supply and low complexity where the "market" can correct itself. After all, if a supplier does not perform, simply rebid the work.

However, as organizations have begun to outsource and procure more complex goods and services, this logic no longer works. All too often buyers become co-dependent on suppliers, switching costs are high, and suppliers have a "locked-in" position.

Oliver E. Williamson, professor of economics at the University of California, Berkeley, has challenged the concept that sourcing is a "make versus buy" decision with his work in Transaction Cost Economics (TCE).[7] Williamson received the Nobel Prize for his work in 2009. One of Williamson's key lessons is that organizations should view sourcing as a continuum rather than a simple market-based make versus buy decision.

Perhaps the best way to think of Williamson's work is to consider (see Figure 9-1) free-market forces on one side and what Williamson refers to as "corporate hierarchies" on the other. In the middle, Williamson advocated that organizations should use a "hybrid" approach for complex contracts.

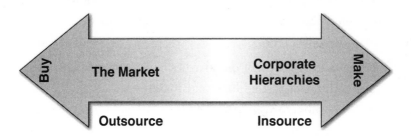

Figure 9-1 A Continuum of Outsourcing Solutions

Developing Corporate Hierarchies: "Make" or Insource

Organizations have long debated the merit of making versus buying goods and services. The industrial revolution enabled corporations to capitalize on big business; the result was vertically integrated companies designed to build and leverage their power. For the most part, big business was met with big government. The default was for organizations to "make" versus "buy" the goods and services they needed to sustain themselves, whether in the private or public sector. The result? Large, powerful organizations that many would classify as "bureaucratic."

The make versus buy debate got a boost when well-respected management visionary Peter Drucker encouraged CEOs to "sell the mailroom" in his 1989 *Wall Street Journal* article. Big thinkers Peters and Waterman started the conversation in their best-selling book *In Search of Excellence* in 1982.[8] By 1990, scholars C. K. Prahalad and Gary Hamel had taken the debate to a new level in their pioneering and popular *Harvard Business Review* article titled "The Core Competence of the Corporation," which encouraged organizations to evaluate their "core competencies."[9] Their message? Most organizations cannot focus on more than five or six core competencies.

Logic prevailed, and CEOs and government officials the world over began to shed internal assets. Goods and services that were customarily controlled in-house, such as information technology, call center/customer care, supply chain services, and back-office finance functions were outsourced.

Government agencies got on board as well. Water Boards began to outsource the design, building, and maintenance of water treatment facilities. Public/private partnerships emerged for public works projects. Defense agencies began buying weapons systems maintenance capabilities instead of simply buying spare parts.

The result was a steady shift to the procurement of more and more complex goods and services.

Scholarly work into the make versus buy debate has brought further clarity. Williamson's work on Transaction Cost Economics (TCE) puts math behind Peter Drucker's commonsense approach. TCE points out that many hidden transaction costs are associated with performing work that is noncore to the organization. One of the downfalls is that when work is performed by an organization's internal resources, there is no competition; this provides little incentive to drive inherent improvements in cost and quality. There is also high administrative control and a legal system that is "deferential to the management." As a consequence, innovations that might come from the market or third parties are not shared or developed as rapidly as management typically likes—if at all.

Because these are additional bureaucratic costs, Williamson advises, "The internal organization is usually thought of as the organization of last resort." In other words, if at all possible, organizations should not invest in developing goods and services that are "noncore."

Using the Market: "Buy" or Outsource

Organizations that choose to procure goods or services typically use the "market" to buy goods and services. The market uses the conventional free market economy for determining how organizations will do business, including establishing a price. A market mode assumes that free market forces incentivize suppliers to compete on low cost and high service. This approach also features an absence of dependency; if buyers or suppliers are not happy, they can switch at any time with relative ease. Governance of the supply base is typically accomplished by switching suppliers or customers if a better opportunity comes along. As a result, the market approach can rely purely on classical contract law and requires little administrative control.

The big advantage to using the market is its simplicity. The market mode enables a competitive process to determine whether an organization is getting a good transaction price. Thus, at its heart is a transactional business model. Competitive bidding processes establish market prices for everything from a per unit price for a spare part, price per call for technical support, price per pallet stored in a warehouse, and price per hour for a janitor to clean a building.

The downside to the market mode is that it often assumes that the purchase is somewhat standardized and therefore available from a variety of suppliers. Consequently, suppliers often "compete" in contracts that pose unnecessary risks.

For example, Williamson notes that service providers might have "specialized investments" that can easily expose the business to significant loss if the contract fails and for which no safeguards have been provided.

One form of specialized investment is in innovations that create value for the buying organizations, such as asset-specific product and process improvements designed to create competitive advantages for the buyer. As suppliers make specialized investments to support innovation, they look at risk versus reward. Often they raise prices to reflect their increased level of risk. However, buyers naturally want reduced prices as well as the benefits of the innovation. Buyers and suppliers often find themselves in a "give and take" as a normal part of market-based negotiations and suppliers seek to develop contractual safeguards. Williamson's research shows that using the market for more complex contracts drives up transaction costs, arguing that complex contracts should use what he refers to as a "hybrid" approach. This hybrid approach embraces a conscious decision to build more trusting and secure supplier relationships in order to drive out opportunism and insert efficiencies in the buyer-supplier relationship.

Hybrid Relationships

Capitalism shows us that the best results come from competition. On the surface this makes sense; it's hard to argue with 200 years of progress under Adam Smith's teachings. However, big thinkers are challenging the concept that highly competitive market forces automatically mean the best result. Transaction Cost Economics and Game Theory indicate there is a better way.

Although the market and hierarchical approaches offer advantages, they also have clear disadvantages. Williamson points out that the market doesn't always work as efficiently as theory would lead one to believe. And buyers may find they don't have skills or money to invest in certain competencies. Game theory teaches us to view a problem through a different lens: one designed to optimize for the problem under review. Every day, more and more research is proving that collaborative, not competitive, games yield consistently better results.[10]

Unfortunately our procurement tools are designed to promote commoditization and competition. This can put a buyer in a Catch-22: organizations want to drive innovation and create a competitive advantage, yet still want to rely on the lowest-cost supplier for a particular good or service. Why is there a Catch-22? Organizations are saying they want a "strategic" supplier, but the nature of how they buy and contract tells a different story—one of commoditization and competition.

Procurement professionals are taught to commoditize goods and services, using Kraljic's "leverage" and "exploit" techniques to help them increase their buying power.[11] They are encouraged to have uniformly available goods and services (for example, commodities) where a buyer can easily compare "apples to apples" and avoid potential supplier protests due to subjectivity in the supplier selection process. In some cases, public procurement policies are even written in such a way that buyers are wary to have one-on-one discussions with suppliers. How can a buyer learn of innovations if it is limited in communicating with suppliers about potential innovative solutions?

To make matters worse, it is not uncommon that procurement professionals are measured on (and often incentivized on!) driving cost reductions through a Purchase Price Variance metric.[12] This drives short-term emphasis on "price" paid versus overall value or a focus on reducing total ownership costs. To top it off, far too many lawyers hunker down with the single-minded goal to shift risk and emphasize shorter-term contracts to limit supplier dependency.

These practices are magnified when combined with a conventional transactional business model where a supplier is paid for every activity. A transactional model pits buyer against supplier with conflicting goals. The more hours, the more units, the more calls, or lines of code written—the more revenue and profit for a supplier. Buyers find their suppliers meet contractual obligations and service level, but they do not drive innovations and efficiencies at the pace the organization wishes. Suppliers argue that investing in their customers' business is risky because buyers will take their ideas and competitively bid the work.

On one hand, organizations want suppliers to close gaps when they lack core competency, wanting suppliers to be innovative and provide solutions. On the other hand, they drive competition and commoditization, preventing suppliers from wanting to invest in innovation. The result is an industry at a crossroads, with both buyers and suppliers wanting innovation but neither wanting to make the necessary investments.

What's needed? Alignment and collaboration between the buyer and the supplier on the sourcing business model that's the right fit for them.

Seven Sourcing Business Models

There is increasing awareness that transaction-based approaches do not always give buyers and suppliers the results they seek. University of Tennessee research and industry-specific experiences applying alternative output and outcome-based approaches for complex contracts demonstrate that alternative sourcing business models are highly viable approaches to the conventional transactional methods.[13] Both output and outcome-based approaches are gaining momentum as senior leaders see positive results from carefully crafted collaborative agreements.

The book *Strategic Sourcing in the New Economy: Harnessing the Potential of Sourcing Business Models in Modern Procurement* provides an in-depth review of seven sourcing business models falling into the three categories along Oliver Williamson's sourcing continuum.[14] This chapter provides the highlights of each model.

- Transactional (Williamson's "Market" category)
  - Basic Provider Model
  - Approved Provider Model
- Relational (Williamson's "Hybrid" category)
  - Preferred Provider Model
  - Performance-Based / Managed Services Model
  - Vested Business Model
- Investment (Williamson's "Hierarchy" category)
  - Shared Service Model
  - Equity Partnerships (for example, joint ventures)

Each sourcing business model differs from a risk/reward perspective and should be evaluated in the context of the goods or services being procured. The characteristics and attributes for each of these approaches are reviewed in detail in the next section. Figure 9-2 shows how the sourcing business models are placed along the sourcing continuum.

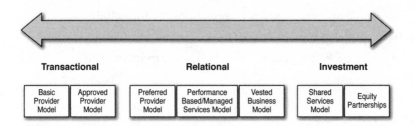

Figure 9-2 Sourcing Continuum

Basic Provider Model

A basic provider model uses a transaction-based model, meaning there is usually a set price for individual products and services for which there is a wide range of standard market options. Typically, these products or services are readily available, with little differentiation in what is offered.

A basic provider model is used to buy low-cost, standardized goods and services in a market where there are many suppliers, and switching suppliers has little or no impact on the business. Buyers typically use frequent competitive bidding (often with pre-established electronic auction calendar events). Often a purchase requisition triggers transactions that signal that the buying company agrees to buy preset quantities of goods or tasks (for example, widgets or hours). Some organizations even use purchase cards for these types of simple purchases.

The buyer-supplier relationship is based largely on a review of performance against basic criteria. For example, did the supplier work the hours claimed? Did the goods received meet the agreed-to quantity, cost, and delivery times?

Basic Provider Example

A hospitality organization with several properties purchased a variety of low-cost basic food items, such as salt, mustard and other condiments, snack items, and pasta. Each property did its own purchasing, and no specific requirements were applied to these basic food items because all items were standard in the marketplace and a number of suppliers provided the products. However, when the organization investigated the number of items that were being procured as basic food items, the estimated number exceeded 16,000 items and multimillions of dollars of annual spend.

The organization believed there was a better way to manage these items. The organization sought to put in place a process that would obtain more detailed information across all properties on these items, without adding resources to manage them and to obtain the lowest market price. The organization implemented a standard e-auction tool that was used by all properties.

Item requirements were entered into the online e-auction tool, the suppliers in the marketplace placed their bids, and the lowest pricing supplier won the order. No negotiations were conducted. A purchase order was generated using standard terms and conditions and distribution program, and the properties exerted limited effort to manage a multimillion dollar spend, which allowed their purchasing resources to focus on higher-cost items.

Approved Provider Transaction Model

An approved provider model uses a transaction-based model where goods and services are purchased from prequalified suppliers that meet certain performance or other selection criteria. Frequently, an organization has a limited number of preapproved suppliers for various spend categories from which buyers or business units can choose. Multiple suppliers mean costs are competitive, and one firm can easily be replaced with another if the supplier fails to meet performance standards.

An approved provider is identified as a prequalified option in the pool of basic providers. Approved providers fulfill preconditions for specified service through a set of criteria or previous experience with performance reliability. To reach approved status, suppliers frequently offer some level of differentiation from other transactional suppliers and provide a cost or efficiency advantage for the buyer. The differentiation could come in the form of geographical location advantage, a cost or quality advantage, or a minority-owned business and is ultimately "approved" to assist with meeting an organization's social responsibility goals.

Procurement professionals often turn to approved providers as regularly solicited sources of supply when bidding is conducted. An approved provider may or may not operate under a Master Agreement, which is an overarching contract with the buying organization. Approved providers may or may not also have volume thresholds to be in an "approved" status. In addition, approved providers may or may not participate in supplier management reviews.

To create a seamless and readily accessible supply chain, many organizations develop lists of approved providers. The advantages are many; a preapproved list saves time when seeking particular goods and services, for example. The approval process ensures parity between bidding qualified suppliers. As an organization selects its approved provider list, it molds the required qualifications to its unique business objectives and strategy.

Approved Provider Example

Intel[15] uses approved providers as part of its Supplier Development Program (SDP), which identifies and confirms that all bidding suppliers are at parity. For that reason, Intel feels confidence in its field of competitors. When it is time for the bid process, Intel can select the lowest-cost supplier without concern about the supplier's capabilities. Intel knows the supplier can meet its needs. In essence, Intel works very hard to commoditize what it is buying to drive pricing competition in the market.

Consider Intel's transportation category. First, Intel rates the capabilities of suppliers that serve the transportation category (DHL, UPS, Expeditors, and so on). Next, Intel works with suppliers to ensure they close any identified capabilities gaps. Intel then rates the suppliers again to confirm that they meet capability standards. Finally, when Intel is ready to seek a supplier, there are typically three capable, preapproved vendors from which to choose. All offer a "standard" service offering. Intel's SDP is a solid strategy for commoditized requirements where there are multiple, interchangeable sources of supply. By ensuring adequate competition, Intel is assured it uses the market to get the lowest possible price.

Preferred Provider Model

A preferred provider model also uses a transaction-based economic model. A key difference between a preferred provider and the other transaction-based models is that the buyer has chosen to move to a more strategic relational model. Thus, contracts with specifically chosen supplier(s) assume a more collaborative relationship. Repeat business and longer-term and/or renewable contracts are the norm.

Similar to an approved provider model, buyers seek to do business with preferred providers to streamline their buying processes. Buying organizations typically enter into multiyear contracts using master agreements that allow them to conduct repeat business efficiently. Preferred providers are still engaged in transaction-based economic models. However, the nature and efficiencies of how the organizations work together go beyond a simple purchase order and begin to consider how a supplier can provide value-added services.

A preferred provider is a supplier that is "prequalified." Often it has unique differentiators—offering value-added services and/or demonstrating acceptable levels of performance. For example, a preferred provider may have a superior software system that interfaces with an organization's own system. Sometimes a preferred provider is chosen because of its high-quality workforce and difficult-to-duplicate expertise. Typical conditions for supplier down-selection of a preferred provider are the following:

- Previous experience
- Supplier performance rating (if the buying organization has a rating system)
- Previous contract compliance performance
- Evidence of an external certification (such as ISO certification)
- Additional contributions to control costs, such as inventory management, training resources, and aligned geographical positioning

It is common for preferred providers to work under blanket purchase orders (POs) and rate cards that make conducting repeat business easy. For example, a labor-staffing firm may have a rate card that lists the hourly rate set for various types of staffing needs. The buying organization can easily request staffing support from the preferred provider using the predetermined blanket purchase orders and rate cards.

Preferred Provider Example

In the Microsoft Preferred Supplier Program (MPSP), suppliers are divided into two distinct levels: Premier suppliers and Preferred suppliers. Preferred and Premier suppliers are a small subset of Microsoft's overall list of approved suppliers referred to as the Approved Supplier List (ASL). Premier suppliers are the featured supplier by category in Microsoft's e-procurement system, meaning that when an employee seeks to buy goods or services, the Premier suppliers are the *recommended* source by Microsoft's procurement organization. This leads to substantial revenue increases when business units or employees "buy" products or services using the procurement group's recommendation.

Microsoft's Preferred and Premier suppliers also enjoy added benefits. Microsoft issues invitations to special events during which Microsoft executives share insights and strategies. Premier suppliers also have access to Microsoft Executive briefings. It is not easy to become a Preferred or Premier supplier at Microsoft.[16]

Performance-Based/Managed Services Model

A performance-based model is generally a formal, longer-term supplier agreement that combines a relational contracting model with an output-based economic model. A performance-based model seeks to drive supplier accountability for output-based service-level agreements (SLAs) and/or cost reduction targets. A performance-based agreement typically creates incentives (or penalties) for hitting (or missing) performance targets.

Sourcing decisions are based not only on a supplier's ability to provide a good or service at a competitive cost but also on its ability to drive improvements based on its core competencies. Performance-based agreements shift thinking away from activities to predefined *outputs* or events. Some organizations call the results outcomes. However, it is important to understand that a performance-based agreement should hold a supplier accountable only for what is under its control. For that reason, in performance-based models, the word *outcome* typically means a supplier's "output." An output is a well-defined and easily measured event or a deliverable that is typically finite in nature. Performance-based agreements require a higher level of collaboration than preferred provider contracts because typically there is a higher degree of integration between a supplier and a buying organization. In addition, buyers need to apply more formalized supplier relationship management efforts to review performance against objectives and specify the incentive or service credit (also referred to as a malice payment or penalty) payments that are embedded in the contracts.

Some service industries are seeing an evolution in managed services agreements where a supplier guarantees a fixed fee with a pre-agreed price reduction target (for example, a 3 percent year-over-year price decrease). The assumption is that the supplier will deliver on productivity targets. These guaranteed savings are often referred to as a *glidepath* because there is an annual price reduction over time. Managed services agreements are a form of a performance-based sourcing business model.

Performance-Based Example

The United States Navy set out to improve the performance of the H-60 FLIR system, which enables the Navy's H-60 helicopter to detect, track, classify, identify, and attack targets such as fast-moving patrol boats or mine-laying craft. When first developed, the FLIR was expected to have at least 500 hours of operation before failure, but in reality was averaging less than 100 hours.

The Navy and Raytheon implemented a 10-year, fixed-price agreement that was priced per flight hour and valued at $123 million. This fixed price by flight hour contract gave Raytheon incentive to improve reliability and help reduce the necessity for removal of these units from the aircraft.

Raytheon also implemented an online Maintenance Management Information System that allowed for real-time data collection by NADEP Jacksonville; an online manual has eliminated the need to have printed copies made and distributed.

In the first three years of the contract, the H-60 FLIR components experienced a 100 percent availability rate and achieved a 40 percent growth in system reliability improvement, as well as a 65 percent improvement in repair response time. Originally, cost savings were projected to be around $31 million but exceed $42 million after just three years. The Navy was recognized by the United States

Secretary of Defense for its "Performance-Based Logistics" contract with Raytheon for its H-60 FLIR program.[17]

Vested Sourcing Business Model

The Vested Sourcing Business Model is a hybrid relationship that combines an outcome-based economic model with a relational contracting model. It incorporates the Nobel Prize–winning concepts of behavioral economics and the principles of shared value.[18] Using these concepts, companies enter into highly collaborative arrangements designed to create and share value for buyers and suppliers above and beyond the conventional buy-sell economics of a transaction-based agreement. In short, the buyer and the supplier are equally committed (Vested) to a joint outcome: each other's economic success over the long term.

A good example is Microsoft and Accenture's multiyear agreement, in which Microsoft challenged Accenture to transform Microsoft's back-office finance operation processes. The agreement is structured so that the more successful Accenture is at achieving Microsoft's goals, the more successful Accenture itself becomes.[19]

The Vested business model was popularized when University of Tennessee researchers coined the term after studying highly successful buyer-supplier relationships. A Vested business model is best used when an organization has transformational and/or innovation objectives that it cannot achieve by itself or by using conventional transactional sourcing business models (Basic Provider, Approved Provider, Preferred Provider) or even a Performance-Based agreement.

These transformational or innovation objectives are referred to as Desired Outcomes. A Desired Outcome is a measurable strategic business objective that focuses on what will be accomplished as a result of the work performed. Desired Outcomes are not a task-oriented service-level agreement (SLA) such as those typically outlined in Preferred Provider or Performance-Based agreements. Rather, Desired Outcomes are strategic in nature and often can be achieved only with a high degree of collaboration between the buyer and provider and/or with investment by the supplier.

Desired Outcomes form the basis of a Vested relationship because the supplier is rewarded for helping the buyer achieve mutually defined Desired Outcomes—even when some of the accountability is shared with the buying organization. Desired Outcomes are generally categorized as an improvement to cost, schedule, market share, revenue, customer service levels, or overall business performance.

Vested Business Model Example

In 2003, Procter & Gamble signed a contract with Jones Lang LaSalle spanning 60 countries and including facility management, project management, and strategic occupancy services. P&G wanted an outsourcing relationship that challenged JLL to not just take *care* of its buildings, but to take *charge* of the buildings.[20]

P&G was clear: the real reason it outsourced was to drive transformation and achieve "the power of AND." Its contracting approach motivated JLL to bring new ideas and determine the best way to get results. P&G shifted the economics of outsourcing to an outcome-based approach, whereby P&G bought Desired Outcomes, not individual transactions or service levels. P&G paid JLL based on its ability to achieve mutually agreed outcomes. P&G and JLL shared an interest in achieving P&G's strategic goals.

A key component for P&G was to focus on the *what*, not the *how*. After P&G decided it was serious about trusting and delegating responsibility to JLL, contract negotiation was considerably simplified: it was JLL's job to figure out what was needed and how to get it done.

P&G and JLL know that measurement drives behavior. Instead of focusing on time and tasks, they focus on measuring success against

P&G's business priorities. Formal governance mechanisms allowed P&G and JLL to refocus priorities as needed. The parties went on to develop a pricing model that incentivized JLL to achieve the Desired Outcomes and drive innovation.

Another important aspect of the P&G/JLL governance structure is that the companies live (and manage) the business following an insight—not an oversight—governance structure. They do this with what they call a "2 in a Box" approach that identifies both a P&G and JLL person as owners of a core process. This assures business plans and action plans are aligned between P&G and JLL.

JLL went from being a new P&G supplier to winning "Supplier of the Year" two years in a row among a field of 80,000 suppliers. P&G is on record saying that Global Business Services has reduced cost as a percentage of sales by 33 percent for its outsourced operations. JLL exceeded the satisfaction target for six consecutive years. JLL was a winner, too, expanding capabilities, profitability, and earning additional workscope during contract renewal.

Shared Services Model

Organizations struggling to meet complex business requirements with a supplier can always invest to develop capabilities themselves (or insource). One approach is to develop an internal shared service organization (SSO) with the goal of centralizing and standardizing operations that improve operational efficiencies. A shared services model is typically an internal organization based on an arm's-length outsourcing arrangement. Using this approach, processes are often centralized into an SSO that charges business units or users for the services they use. In some instances, SSOs are formed externally to the company (such as a subsidiary).

SSOs typically act like outsourced suppliers, performing services and then "charging" their internal customers on a per-transaction or actual cost basis. Generally, SSOs mirror conventional preferred provider models. The main difference is that the SSO is an internal supplier rather than an external supplier.

Organizations can use a shared services model for a variety of functional services, such as human resources (HR), finance operations, or administrative services (such as claims processing in health care). For example, large organizations may centralize HR administration into an SSO to provide benefits management to their own employees and even external clients. Small enterprises can benefit from a shared services model by joining forces to create specialized service centers that economically provide a functional service to each of the smaller firms.

Shared Services Example

In 1995, Bell Canada's distribution operations were operating at service levels at 10 to 15 percent below industry average and at a cost base of $100 million. Bell Canada (the largest telecom services organization in Canada) decided to spin off the assets and the staff of the distribution business into a standalone, wholly owned subsidiary known as Progistix Solutions Inc. (PSI). The idea was that by creating a separate shared services entity with its own P&L, PSI would be driven to operate more efficiently. PSI was chartered to provide a full range of order management and inventory management business processes for all of Bell's operating businesses, and a new CEO was brought in to turn around the business.

At its inception, PSI had an estimated revenue stream (benchmarked by Deloitte) of $55 million against its cost base of $100 million. Progistix had a mandate to achieve a financial breakeven state and to meet industry average service levels.

With its own P&L, the shared services group carefully reviewed where it needed to invest in business processes and technology to meet its charter of becoming a profitable business unit and raising service levels to its Bell counterparts. PSI invested in three key areas:

- Replaced the aged technology infrastructure and outdated applications

- Renegotiated the four collective agreements to align wage rates and work rules with the logistics services market

- Commenced the long process of culture change from an entitlement-based telecom services organization to a market-focused logistics services competitor

Clearly, the cultural change would be the most difficult. By moving noncore functions to an organization dedicated to enhancing quality in its respective field (shared services or outsourcing), these employees gain respect and self-confidence, enabling them to perform at much higher levels.

In addition to the attention to the preceding key priorities, the management team was driven through profit-sharing incentives to dramatically reduce costs in all parts of the organization. As a result of its efforts, PSI reduced its costs by $45 million, yielding a breakeven position in 1998. In addition, systematic improvements raised service levels to industry standards, and more than 95 percent of the orders processed during the day were picked, packed, shipped, and delivered to customers by the end of the next day.

During the next two years, PSI was able to generate industry standard profits and grow revenues by 15 percent.

Equity Partnerships

Equity Partnerships are legally binding entities. They take different legal forms, from buying a supplier (an acquisition), to creating a subsidiary, to equity-sharing joint ventures or cooperative (co-op) arrangements. Equity Partnerships are best used when an organization does not have adequate internal capabilities and does not want to outsource.

Some organizations decide they do not have internal capabilities and do not want to invest in a shared services organization. In these cases, organizations may opt to develop an Equity Partnership such as a joint venture or other legal form in an effort to acquire mission-critical goods and services.

Equity Partnerships, by definition, bring costs "in house" and create a fixed cost burden. As a result, Equity Partnerships often conflict with the desires of many organizations to create more variable and flexible cost structures on their balance sheet.

Joint Venture Example

The consumer electronics giants Samsung Electronics and Sony established a 50–50 joint venture in 2004 for the production of liquid-crystal displays for flat panel televisions. The companies formed a new organization, S-LCD Corp., near Seoul, South Korea, with an initial capital budget of nearly $2 billion.

The two tech giants—and fierce industry rivals—structured the venture so that stocks in S-LCD were held by South Korea's Samsung at 50 percent plus one share of stock and 50 percent minus one by Japan's Sony. "The two organizations will invest evenly, but Samsung has the ultimate initiative," said a Sony spokeswoman.[21]

Upstart Samsung had begun construction of an LCD production facility in 2003 at a large projected capital expenditure for what was then a relatively new technology and market. Sony had no production base for large LCD panels. Joint collaboration was thus advantageous for both organizations. The deal was also controversial: Sony had pulled out of a Japanese-state-backed LCD-panel development group to close the deal with Samsung.

In 2006, *Bloomberg Business Week* described the venture as a win-win: "They have pulled off one of the most interesting and fruitful collaborations in global high-tech by jointly producing liquid-crystal display (LCD) panels. And it's an alliance that is reshaping the industry."[22] The venture was instrumental in Sony's introduction of the hugely successful Bravia LCD-TV lineup. It also put Samsung's own LCD-TV business on the map as a trendsetter in the LCD-panel industry, aided by Sony technology that helped ensure high-quality, sharp TV pictures. "The Sony-Samsung alliance is certainly a winwin," said Lee Sang Wan, president of Samsung's LCD unit.

The alliance had industrywide impact in the TV market for large-screen sets. It also changed the pecking order among LCD-TV makers. In 2008 the organizations strengthened the venture by committing another $2 billion to build a new facility to produce so-called eighth-generation panels. In the intervening years, despite global economic and financial turmoil, currency fluctuations, heavy competition, and new entrants in the LCD and electronics market, and more recently the earthquake and tsunami in Japan, the S-LCD venture has survived.

The earthquake and faltering global demand in the LCD market did force S-LCD to reduce capital by $555 million in April 2011. There were even rumors that the joint venture might be dropped because of losses in Sony's TV business, but Sony quashed that idea. "Televisions are a core business for Sony and it would be unthinkable for us to shrink that business," said Kazuo Hirai, Sony's executive deputy president. The venture is both unusual and remarkable in terms of its scope and duration. Two fierce competitors put their rivalry aside to achieve the win-win in an emerging market.

Different Models Need Different Systems

While business needs have evolved, the fundamental nature of how goods and services are procured has not. The vast majority of organizations (public and private) still use the same transaction-based approach for procuring complex goods and services as they do to buy more simple commodities and supplies.

Unfortunately, many business professionals wrongly assume that a transaction-based business model is the only way to architect a supplier contract. *For simple transactions with abundant supply and low complexity, a transaction-based business model is the most efficient model.*

The real weakness of a transaction-based approach emerges when any level of complexity, variability, mutual dependency, or customized assets or processes is part of the transaction. This is because the transactional approach cannot produce perfect market-based price equilibrium in variable or multidimensional business agreements. In many instances, hybrid sourcing business models built with relational contracts and output or outcome-based economic models are more appropriate.

Think of a sourcing business model as a "system," because each is purpose-built to optimize the business needs given critical operating factors. Each situation is different: there is no one-size-fits-all sourcing business model.

For this reason it is vital that organizations work together to choose the most appropriate sourcing business model for their situation.

Supplier Integration via Vested Relationships

The Vested Sourcing Business Model is gaining increasing attention as organizations such as Procter & Gamble, Dell, Intel, Novartis, TD Bank, and others begin to share their knowledge about how a Vested model can drive innovation.[23, 24, 25, 26, 27]

UT researchers codified the Vested methodology with the strategic intent to help teach organizations how to harness the power of highly collaborative supplier relationships. Five books have been published on the Vested model, and UT offers six courses as part of its Certified Deal Architect Program. The Vested Sourcing Business Model is a combination of outcome-based and shared-value principles that is best used when a company wants to move beyond having a service provider perform a strict set of directed, transaction-based tasks and instead develop a collaborative solution based on mutual advantage to achieve desired outcomes.

The Vested methodology is anchored around five "rules." The Five Rules increase innovation and improve efficiency by moving away from the conventional, transaction-based and risk-averse approach to contracting. The Five Rules are as follows:

1. *Focus on outcomes, not transactions.* Flip the thinking from a focus on specific transactions to desired outcomes; instead of buying transactions, buy outcomes, which can include targets for availability, reliability, revenue generation, employee or customer satisfaction, and the like.

2. *Focus on the what, not the how.* If a partnership is truly outcome-based, it can no longer have a multiplicity of service-level agreements (SLAs) that the buyer is micromanaging. The outsource provider has won the contract because the provider is supposed to have the expertise that the buyer lacks. So the buyer has to trust the supplier to solve problems.

3. *Have clearly defined and measurable outcomes.* Make sure everyone is clear and on the same page about their desired outcomes. Ideally, there shouldn't be more than about five high-level metrics. All parties need to spend time collaboratively, during the outsourcing process and especially during the contract negotiations, to establish explicit definitions for how relationship success will be measured.

4. *Pricing model with incentives that optimize the business.* Vested does not guarantee higher profits for suppliers; they are taking a calculated risk. But it does provide them with the tools,

autonomy, and authority to make strategic investments in processes that can generate a greater ROI and value over time.

5. *Establish an insight, rather than oversight, governance structure.* A flexible and credible governance framework makes all the rules work in sync. The structure governing an outsource agreement or business relationship should instill transparency and trust about how operations are developing and improving.

Working in conjunction with the Five Rules are Ten Elements that form the basis of a Vested contract. Table 9-1 shows the linkage and interaction between the Rules and the Elements:

Table 9-1 Rules and Elements of Vested Contracts

Rule 1: Outcome-Based Versus Transaction-Based Business Model

Element 1	Business Model Map
Element 2	Shared Vision Statement and Statement of Intent

Rule 2: Focus on the What, not the How

Element 3	Statement of Objectives/Workload Allocation

Rule 3: Clearly Defined and Measurable Desired Outcomes

Element 4	Clearly Defined and Measurable Desired Outcomes
Element 5	Performance Management

Rule 4: Pricing Model Incentives Are Optimized for Cost/Service Trade-offs

Element 6	Pricing Model and Incentives

Rule 5: Insight Versus Oversight Governance Structure

Element 7	Relationship Management
Element 8	Transformation Management
Element 9	Exit Management
Element 10	Special Concerns and External Requirements

An overview of each of the Ten Elements and how they can help companies develop a Vested Outsourcing agreement follows.

Rule #1: Focus on Outcomes, Not Transactions

The first two elements complement the first rule by setting the buyer and supplier on a path that will ensure positive business outcomes for both parties.

Element 1: Business Model Map

This first step is to understand the business at hand and document an outsourcing business model. It is vital to devote time to mapping potential outcomes and to see how well the parties are aligned to each other's goals. Jointly mapping a model will pinpoint the transactions of value between the parties, leading almost inexorably to collaboration, loyalty and mutual satisfaction, market share, and sustainable profit.

Element 2: Shared Vision and Statement of Intent

With the business model understood and mapped, the parties then work together on a joint vision that will guide them for the duration of the Vested relationship. A cooperative and collaborative mindset opens a conversation between the parties: The result is that they share what is needed, admit to gaps in capability, and aim to focus on the benefits that the other party can bring to alleviate any gaps in capability. That vision and alignment form the basis of a Statement of Intent.

Rule #2: Focus on the **What,** *Not the* **How**

The purpose of Rule 2 is to understand and then document the workscope between the buyer and the supplier. The main difference between a conventional approach and the Vested approach is that under the latter approach, the buyer specifies *what* it wants and moves the responsibility of determining *how* the work gets delivered to the supplier/service provider.

Element 3: Statement of Objectives/Workload Allocation

Depending on the scope of the partnership, the company transfers some or all of the activities needed to accomplish agreement goals to the service provider. Together they develop a Statement of Objectives (SOO), which is very different from a standard Statement of Work (SOW). Simply put: an SOO describes intended results, not tasks. Based on the SOO, a service provider will draft a performance work statement that defines in more detail the work to be performed and the results expected from that work.

Rule #3: Clearly Defined and Measurable Outcomes

Element 4: Top-Level Desired Outcomes

To have an effective, successful Vested relationship, the parties must first work together to define and quantify their Desired Outcomes. Outcomes are expressed in terms of a limited set of high-level metrics. It is imperative that the parties spend time during the transition, and particularly during agreement negotiations, to define exactly how relationship success is measured. After the desired outcomes are agreed upon and defined, the supplier proposes a solution that will deliver the required level of performance at a predetermined price.

Element 5: Performance Management

After the Desired Outcomes, Statements of Intent, and SOOs are in place and the agreement is implemented, the parties then measure actual performance to determine if the desired outcomes are achieved. These statements must include high-level performance management measures that are easily understood by business stakeholders and all parties involved in the process. The metrics will help align performance to strategy.

Rule #4: Pricing Model Incentives Are Optimized for Cost/Service Trade-offs

There is no one-size-fits-all Vested pricing model, but the benefits of a fair pricing structure, reached through cooperation, flexibility, and innovative thinking, are obvious. Getting pricing and the pricing relationship right speaks directly to the bottom line of the enterprise and its ultimate success.

Element 6: Pricing Model and Incentives

To attain Desired Outcomes, the parties must have a properly structured pricing model that incorporates incentives for the best cost and service trade-off. The focus of many procurement professionals to outsourcing remains stuck in typical and conventional approaches: getting the lowest possible service and labor pricing. Inherent in the Vested Sourcing Business Model is a reward for service providers to invest in process, service, or associated product that will generate returns in excess of agreement requirements. This element gives service providers the authority and autonomy to make strategic investments in processes and product reliability that can generate a greater return on investment than a conventional cost-plus or fixed-price-per-transaction agreement might yield.

Incentives are a key component of this approach because service providers are taking on risk to generate larger returns on investment. An incentives package delivers the most commercially efficient method of maintaining equitable margins for all parties for the duration of the relationship.

Rule #5: Insight Versus Oversight Governance Structure

A sound governance structure provides consistent management along with cohesive policies, processes, and decision rights that enable parties to work together effectively and collaboratively. A governance framework enables the parties to manage performance and achieve transformational results throughout the life of the agreement. In addition, a good governance structure creates an environment for understanding the business better and to make proactive changes that can help a company control its actions.

Element 7: Relationship Management

A relationship management structure creates joint policies that emphasize the importance of building collaborative working relationships, attitudes, and behaviors. The structure is flexible and provides insight into what is happening with the parties' Desired Outcomes and their relationship. The parties monitor this relationship within the framework of a flexible governance structure that provides top-to-bottom insights into what is happening.

Element 8: Transformation Management

The Vested partnership is a new relationship model that recognizes that business is dynamic in nature and that company ecosystems and business needs may change over time. This element addresses the need to manage transformation, including transitioning from existing approaches—along with change management after the new agreement is up and running. The focus is on continuity and mutual accountability for desired outcomes, end-to-end business metrics, and the creation of an ecosystem that rewards innovation and an agile culture of continuous improvement.

Element 9: Exit Management

Sometimes the best plan simply does not work out or is trumped by unexpected events: business happens, and companies should have a plan when assumptions and conditions change. This governance element provides an exit management strategy and template to handle future unknowns.

Element 10: Special Concerns and External Requirements

The final element recognizes that all agreements are unique and that many companies and service providers must understand and adhere to special requirements and regulatory protocols. The governance framework may need to include additional provisions that address specific market, local, regional, and national requirements. For instance, in supplier and supply chain relationships involving information technology and intellectual property, security concerns may necessitate special governance provisions outside of the normal manufacturer-supplier relationship. Supply chain finance and transportation management are other areas that often require special handling under the governance framework.

Developing and integrating your supply chain using Vested's Five Rules and Ten Elements is much more than delivering a higher level of service on a given activity, a blur of metrics, or simply counting transactions or buying labor more cheaply. It is a *fundamental sourcing business model shift* in how buyers and suppliers can work together for the win-win. This is why organizations such as Dell are turning to Vested as a way to work more strategically with key suppliers such as GENCO—Dell's reverse logistics supplier.

Dell and its strategic partner GENCO followed the rules with great success. The Dell/GENCO relationship was longstanding and was expanded in 2009 when GENCO agreed to acquire Dell's buildings, assets, and people under a three-year outsourcing contract. The problem was that it was a strategic relationship, but the transactional structure of their agreement was far from strategic: it was a typical transaction-based contract in which GENCO assumed the risk of meeting a set "price per activity" while maintaining service levels. The agreement worked reasonably well for a time, but Dell's leaders continued to face cost pressures, and they insisted on an "every dollar, every year" procurement principle—even though under the contract, GENCO assumed much of the risk at the "set price" contract terms.

The seeds were sown for a difficult endgame unless both companies could transform their relationship through trust, collaboration, and the Vested "what's in it for us" mindset. They succeeded by structuring a strategic, outcome-based commercial agreement with true win-win economics.

It was a huge success for Dell's Global Outlet—reducing its cost structure by 32 percent, increasing revenues to record highs, and becoming more environmentally sustainable by reducing the scrap level of old and damaged hardware by 62 percent. GENCO also benefited with a tripling of its margins.

John Coleman, GENCO's general manager, said the Vested approach was a key reason that enabled GENCO to drive innovation for Dell. "It gave us the freedom to get creative. It's like we broke open a new innovation piñata. GENCO employees now know that we will share in the reward for good ideas. Now, every quarter we make new priorities that align with our defined mutual outcomes."

The Dell and GENCO journey proves that the Five Rules provide an alternative to the relentlessly competitive "I-win-you-lose" transaction mindset by creating commercial structures designed to foster a culture of cooperation, trust, and innovation. The buyer and supplier become Vested in each other's success.

Conclusion

A clear message is emerging: the business battlefields of this century will be based on harnessing the power of your suppliers. Tomorrow's winners will no longer play yesterday's competitive win-at-all costs game with key suppliers. The playing field is no longer one of lowest cost or best value, but one of highly collaborative relationships with suppliers that can help drive transformation and innovation in your organization. After all, if firms are going to compete "supply chain to supply chain," shouldn't all the links in the supply chain work together?

Endnotes

1. Kate Vitasek is a Faculty Affiliate for Graduate and Executive Education in the University of Tennessee's Haslam College of Business, and Founder of Supply Chain Visions.

2. Henke, J.W. Jr., T.T. Stallkamp, and S. Yeniyurt. 2014. Lost Supplier Trust,... How Chrysler Missed Out on $24 Billion in Profits Over the Past Twelve Years," *Supply Chain Management Review*. May/June. Available at http://www.ppi1.com/uploads/wri-profit/scmr-lost-trust.pdf.

3. Drucker, P.F. 2005. Sell the Mailroom. *Wall Street Journal*, November 15. Available at http://www.wsj.com/articles/SB113202230063197204.

4. Friedman, T.L. 2005. *The World Is Flat: A Brief History of the Twenty-first Century*. New York, NY: Farrar, Straus & Giroux:.

5. Chick, G. and R. Handfield. 2012. *The Procurement Value Proposition*. London: Kogan Page.

6. Lafley, A.G. and R. Charan. 2008. *The Game-Changer: How You Can Drive Revenue and Profit Growth with Innovation*. New York: Random House, pp. 13–36.

7. Williamson, O.E. 2008. Outsourcing: Transaction Cost Economics and Supply Chain Management. *Journal of Supply Chain Management* (44) 2: 5–16.

8. Peters, T. and R. Waterman. 1982. *In Search of Excellence*. New York: Harper & Row.

9. Prahalad, C.K. and G. Hamel. 1990. The Core Competence of the Corporation. *Harvard Business Review* (68) 3: 79–91.

10. Nobel laureate John Nash's research into Game Theory showed that when individuals and organizations work together to solve a problem, the results are always better than if they had worked separately or played against each other. Similarly, the political science professor Robert Axelrod delved into the nature of competitive relationships through his research into the Prisoner's Dilemma game. His book, *The Evolution of Cooperation*, showed that playing "nice"—or cooperating—led to the best results while maximizing mutual gains for all players.

11. Kraljic, P. 1983. Purchasing Must Become Supply Management. *Harvard Business Review*, September. https://hbr.org/1983/09/purchasing-must-become-supply-management.

12. Purchase Price Variance (PPV) is a procurement metric that measures a procurement organization's (or an individual procurement professional's) effectiveness at meeting cost savings targets. The thinking about PPV is simple: If you spend $1.10 on a widget, then try to get the same widget next year for $1.00. The better the price that's obtained, the better the PPV. Many organizations reward their buyers on a PPV metric.

13. See Vitasek, K., K. Manrodt, and M. Ledyard. 2013. *Vested Outsourcing: Five Rules That Will Transform Outsourcing.* New York: Palgrave Macmillan; Vitasek, K., K. Manrodt, and J. Kling. 2012. *Vested: How P&G, McDonald's, and Microsoft Are Redefining Winning in Business Relationships.* New York: Palgrave Macmillan; Vitaek, K., J. Crawford, and J. Nyden, 2011. *The Vested Outsourcing Manual.* New York: Palgrave Macmillan; and Nyden, J., K. Vitasek, and D. Frydlinger. 2013. *Getting to We: Negotiating Agreements for Highly Collaborative Relationships.* New York: Palgrave Macmillan.

14. Keith, B., K. Vitasek, K. Manrodt, and J. Kling. 2015. *Strategic Sourcing in the New Economy.* New York: Palgrave Macmillan.

15. Vitasek, K. and J. Kling. 2015. *Vested for Success Case Study, The Innovator's Dilemma: How Intel and DHL Drove a Paradigm Shift in Procurement.* University of Tennessee, Office of Business Administration; Teaching Edition.

16. Microsoft shares its procurement program details on its website at www.microsoft.com/en-us/procurement/msp-overview.aspx.

17. Department of the Navy, Commander, Naval Supply Systems Command, Nominations for the Secretary of Defense Performance-Based Logistics Award, June 5, 2005. Also see: The Secretary of Defense Performance-Based Logistics Awards Program for Excellence in Performance-Based Logistics; Summary of Criteria Accomplishments, Section 2, https://acc.dau.mil/adl/en-US/548810/file/68119/PBL%20Award%20Pkg%202006%20Sub-system_H-60%20FLIR.pdf.

18. Behavioral economics is the study of the quantified impact of individual behavior on the decision makers within an organization. The study of behavioral economics is evolving more broadly into the concept of relational economics, which proposes that economic value can be expanded through positive relationships with mutual advantage (win-win) thinking rather than adversarial relationships (win-lose or lose-lose). Shared value thinking involves entities working together to bring innovations that benefit the parties—with a conscious effort that the parties gain (or share) in the rewards.

19. Vitasek et al. *Vested: How P&G, McDonald's, and Microsoft Are Redefining Winning in Business Relationships.*

20. This example is an excerpt from Vitasek et al., *Vested: How P&G, McDonald's, and Microsoft Are Redefining Winning in Business Relationships.* pp. 13–37.

21. Hara, Y. 2004. Samsung, Sony Complete LCD Joint Venture Deal. *EE Times-Asia,* March 11.

22. Samsung and Sony's Win-Win LCD Venture. 2006. *Bloomberg Business Week*, November 28.

23. A longer version of the P&G case study is available in Vitasek et al. *Vested: How P&G, McDonald's, and Microsoft Are Redefining Winning in Business Relationships.*

24. The complete Dell case study is available in the book *Vested Outsourcing*. Vitasek, K., M. Ledyard, and K. Manrodt, 2003. *Vested Outsourcing: Five Rules That Will Transform Outsourcing*, 2nd ed. New York: Palgrave Macmillan. Reproduced with permission of Palgrave Macmillan.

25. The Intel case study is taught at the University of Tennessee as part of the Certified Deal Architect program.

26. Emmanuel Cambresy shared the Novartis case study at the BioPharma outsourcing conference in Munich, Germany, June 10, 2015.

27. Kristi Ferguson (TD Bank) and Anthony Cho (CBRE) shared their success with Vested at the CORENET Global Annual Conference, October 19, 2015.

10

Raw Material Feast or Famine: Integrating Supply Networks to Overcome Resource Scarcity

By John E. Bell and Christopher W. Craighead[1]

Have you ever tried to make a peanut butter and jelly sandwich without bread? Not an easy task! Even if you are "successful," it gets kind of messy during the process. So, too, similar situations can arise when critical raw materials are scarce in the short or long term. These scarce resources can hinder and, in extreme cases, shut down firms and their supply chains and in doing so, be devastating to performance. Unfortunately, there is not a shortage (no pun intended) of examples of resource scarcity encounters.

One well-known supply event occurred in December 2003 when Nissan was caught off guard by a lack of steel from its suppliers. This ultimately caused it to shut down two of its assembly lines for a short period due to a lack of materials. Although the exact dollar impact was not made apparent, shutdowns in the automotive industry are extremely costly (estimates are as high as $200K *per minute* of shutdown). Toyota and Suzuki were also impacted by the same steel supply disruption, which stemmed largely from an unforeseen spike in demand. Prior to this event, there were often surpluses in the steel market, and perhaps the Japanese automakers had become somewhat insulated from the potential risk that surrounded this important commodity.

Later, from 2010 to 2013, other "rare-earth" metals took center stage, whereby high-tech firms had to significantly adjust their supply chains to manage the scarcity of these crucial materials. The production of many products was at stake, including batteries, magnets, lasers, LED lighting, and wind turbines. This supply shortage was driven primarily by the Chinese government's restrictions on the export of rare-earth metals (China was producing more than 90 percent of the world's rare-earth metals at the time). This crisis not only threatened the supply and increased prices for raw materials for many firms such as Honda, GE, and Hitachi, but it also inspired the mining industry to start up new mining operations around the world for rare-earth metals. Today, metals scarcity continues to be an issue. For example, the potential for shortages of nickel seems to be looming—what supply chains will be affected, and how severely? Time will tell.

Such risks are not limited to metals, as revealed by Coca-Cola's disruption stemming from water use in one of its bottling plants in the province of Kerala, India, in 2005. The catalyst of the disruption stemmed from allegations that Coke was damaging the water supply in the area by overusing local freshwater in its bottling operations. Ultimately, Coke was unable to keep a multimillion dollar facility operating because of resulting protests and government sanctions. Unfortunately, this freshwater disruption has been the source of negative media attention for Coke in one of its largest global markets, India, and has led to a number of cost increases. The water problems for Coke seem to continue—in April of 2015, the company suffered a similar setback in India when the government revoked a license to build another plant over water-related issues. Similar problems with freshwater supplies plague companies around the world. Accordingly, companies such as Coke, ABInBev, Nestle, Miller Brewing, Procter & Gamble, Unilever, and General Mills have now all invested heavily into water resource management efforts to ensure their operations have a sustainable supply of water.[2]

Renewable resources such as agricultural crops can also create upstream supply risks for companies and thus be the source of a supply disruption. Starting as early as 2009, the world's supply of cocoa began to be a global issue, and the problem intensified in the main supply locations in Western Africa during 2014. The problems are directly related to the threat of disease in the region (Ebola). The disruption in cocoa supply and increasing prices for cocoa have been devastating to some firms. For example, Mondelez International experienced a 72 percent decrease in its 2014 fourth-quarter profits due to spikes in the prices of cocoa on the world market.[3] Other companies such as Hershey and Nestle have been similarly impacted. The world supply of cocoa is expected to problematic well into 2020.

Supply Risk—It Is There, Whether You Manage It or Not

These stories highlight a growing fact about supply chain management during the past decade: supply chain risks, including the supply of raw materials, are growing. In fact, according to a report from the *Financial Times* in May 2015, supply risks have more than tripled in the past 20 years. They cite an industry index from the Chartered Institute of Procurement and Supply (CIPS) that calculates supply chain risk dating back to 1995. The trend of the index indicates that there is more supply chain risk now than at any other time in the past 20 years. However, the dynamics of supply risks are different by global region. For example, the CIPS index recently improved for North America, but the global rating worsened due to increased supply risks in Asia, Africa, and Eastern Europe.[4]

Accordingly, managing risk has been identified as one of the critical roles carried out by supply chain managers. Although our focus in this chapter lies on understanding supply-side risk and the potential for disruptions in the flow of raw materials to the firm, we believe our proposed approach can be adapted to other aspects of the supply chain. The central goal of our approach is to create continuity (that is, absence of disruption) and enhanced resiliency (the amount of time it takes to get back to a normal—or better—operating level). As such, we will look at the practices of some of the most successful mechanisms that firms have put in place. The lynchpin of creating and maintaining supply continuity and resiliency is *integration*. As such, we will explore how firms integrate with their upstream supply chain network to combat risks. Further, we discuss how firms can build and dynamically adapt these integration capabilities to ever-changing situations that can yield disruptions.

Understanding the Network of Risks

What are companies doing about these increased risks in their supply networks? In the cases of the companies we just described, many of them have taken action to better manage upside supply risk in their supply chain. In fact, companies such as Nestle, GE, Nissan, and Coca-Cola have become leaders in managing supply risk and issues with raw material shortages, and we can learn a great deal from their experiences in dealing with supply risks.

Why is managing supply chain disruptions such an important issue for firms? First, you must understand the complexity and global nature of the supply chains that provide materials to the commercial manufacturing companies that make the majority of the goods and services for the global economy. Fortune 500 firms such as Apple, General Motors, Caterpillar, and almost all the others have built supply networks that source raw materials for their operations from around the globe. However, these networks are often very complex and include thousands of suppliers in an interconnected network that can literally have dozens of layers. Such complexity adds risk to the delivery of goods, making it difficult for firms to even maintain visibility into the actual structure of their supply networks. This fact has been highlighted by the recent inability of many U.S. firms to determine and track how raw materials regulated under Conflict Minerals legislation had even entered their supply chains. It has been reported that 90 percent of the 1,262 firms that filed conflict mineral reports in 2014 were not able to determine if their products were free of conflict minerals.[5] This reveals that many firms cannot fully map their supply chains, or at least do not have the needed transparency into what is happening across the supply networks that provide materials for their finished goods.

Supply risks are further magnified when that supply network has certain physical characteristics that leave it vulnerable to disruption. For example, when suppliers are grouped in close physical proximity to each other, the impact of a disruption increases, and the ability to recover from a disruption event in a dense part of the supply network is lower. Additionally, academic research has shown us that supply chain disruptions from natural disasters, supply failures, and other unexpected natural disasters can directly impact the shareholder value and potentially destroy the long-term viability of the firm. Therefore, senior managers need to be committed to managing supply risks, and they need to invest resources in the operational detection, mitigation, and recovery from disruptions that threaten the supply of materials in their upstream supply chain network.

Managing the Network of Risks: Detection, Mitigation, and Recovery

Supply chain management includes the oversight and management of a firm's upstream and downstream processes that cross company boundaries. This includes directing the flow of materials and related information across a wide array of suppliers and customers that produce goods and services for end customers in the marketplace. The number of suppliers and customers and their connection to each other can vary widely from product to product. Therefore, a supply chain for any particular product should be thought of as a complex physical and social network with a high number of links between firms, who all have different operational and strategic objectives and varying strengths and weaknesses. However, although many companies do not focus much beyond their first-tier suppliers, the success of a firm's supply chain may often depend upon suppliers three or four layers back in the network who provide basic raw materials to a variety of industries and the competing firms in those industries. We believe that such lack of awareness about the upstream supply network can leave a firm highly vulnerable to supply risks and the potential for disruption in the flow of key raw materials for its own operations.

Detection

Managing disruptions resides with the firms' ability to detect either pending or realized supply disruptions. Yet, upstream disruptions in a network are not isolated to a single level and can trickle through multiple layers of the supply chain. However, most firms focus more on downstream activities in their networks to keep their customers satisfied. In fact, it is believed that many firms rarely look beyond one level back in the supply side of their supply chain. This can be a real problem for firms when a disruption occurs multiple layers back in the supply network. For example, in the Japanese earthquake and tsunami in 2011, it was second- and third-tier silicon wafer and semiconductor manufacturers that were disrupted, and this caused some electronics and automotive supply chains to be impacted for more than six months. At first observation, many of the larger computer and electronics firms did not believe they would be impacted by the event, because none of their first-tier suppliers were located in the impacted region of Japan, but later they realized that it was their suppliers' suppliers that were disrupted. Therefore, the detection of disruptions must take a broader view of the entire supply chain and look for locations and regions in the network where risks are the highest. In addition, a strategy for detection is needed to allocate limited management resources to monitor such high-risk areas in the network to more quickly detect and disseminate information about the disruption to the rest of the network—similar to monitoring methods suggested by scholars Lambert and Cooper.[6]

As firms map their supply chains, it quickly becomes evident that they often share network nodes (suppliers) with their competitors, and that supply chains for competing products are not completely separate from each other. For example, many of the same raw material and component suppliers may be a supply source for competing electronics manufacturers such as Dell, Apple, and Hewlett-Packard. These overlaps create unique competitive situations in the upstream supply chain, and in extreme cases, some firms may try to actively interdict or disrupt the supply sources of their competitors. Therefore, it is clear that firms need to invest resources in monitoring important nodes and links in multiple layers of the upstream supply chain in order to detect potential disruptions. However, it has also been pointed out that some links in the network are more important to manage than others. For example, when a single source of supply exists, or when a supplier provides a large percentage of the final value of a product, the firm may wisely decide to focus more resources on monitoring that particular part of the network. Unfortunately, supply networks for industrial firms can be expansive, and only a limited amount of management resources is available to commit to detection. Therefore, a firm should have a cost-effective strategy for monitoring and detecting disruptions in its supply chain.

Mitigation

Mitigation involves lessening the negative effects of the disruption. Traditional mitigation tactics for risk and disruptions often included adding alternative suppliers, additional manufacturing capacity, and additional levels of inventory throughout the nodes of a supply chain network. However, in today's competitive world, the costs of such capacity and inventory are often prohibitive for a firm. In fact, efficiency initiatives and supply chain strategies in the past two decades, such as CPFR, Lean Management, and Value Stream Mapping, often lead firms to substantive changes that have eliminated much of the buffer inventory, nonvalue added time, and excess labor and equipment (capacity) from the supply chain.

More recent methods to mitigate supply chain disruptions have focused more on the idea of building flexibility and standardization across the supply chain to react more quickly to a variety of events in the uncertain future. The reduction of lead times through parallel processing, flexible tooling, and reduction of setup times in manufacturing has greatly improved operational flexibility. Similarly, improved transportation methods, standardization of logistics processes and equipment across geographic locations, and collaborative supplier relationships have helped to create additional flexibility needed to respond to a disruption somewhere in the network. However, in the face of truly impactful disruptions, these tactics may not fully mitigate the impact of a disruption; therefore, firms will still be challenged to recover from disruptions.

Recovery

The ability to quickly recover from a disruption is an important third step in managing supply chain disruptions. While detection and mitigation are used to improve the continuity of firm operations, the ability to recover from the impact of a disruption is created by the resiliency of the supply chain. Methods such as redundancy and hedging of supply sources are ways to build resiliency and ensure long-term survival of the firm. Such methods are usually more costly in the short run, but the resulting resiliency helps a firm survive and recover from a major disruption and return to its previous operating capabilities and output. Academic authors Blackhurst et al.[7] have previously described a number of other characteristics that make a firm's supply chain more resilient, including the use of select human, organizational, and physical resources. For example, having employees in the firm who know how to complete postdisruption analyses can greatly influence the ability to recover from disruption events. Similarly, having predefined or self-executing contingency plans that can be effectively implemented is an important resiliency capability for responding and recovering from a disruption, as is the ability to quickly redesign the physical supply chain.

Disruption events and the resulting recovery management effort can also impact the future relationships between firms, and therefore may increase or decrease the future level of integration needed between firms. Some researchers have described how trust between firms can be positively and negatively influenced in the occurrence and recovery from a supply chain disruption, which we posit could then impact the ability to manage future disruptions.[8] Additionally, a major supply disruption could also lead to the recognition of a new type of threat in the supply chain or could lead to the recognition by partner firms that a particular region of the supply chain network is facing heightened risks. Such observations during the recovery phase of a disruption can subsequently lead firms to rethink the integrative activities they have with other firms in a particular part of the supply chain network. More strategically, it could lead to decisions to dedicate more managerial resources and increase collaborative activities with partner firms in order to shore up weaker aspects of the supply chain.

Integration as an Enabler of Continuity and Resiliency

Unfortunately, some businesses struggle to integrate and stream-line decision making across their own internal business functions. Yet, we believe that firms not only need to be integrating internally, but they also need to integrate with external supply chain partners. This external integration will allow them to detect, mitigate, and recover from potential disruptions in the flow of raw materials. But what is integration?

Academic scholars Esper et al.[9] discuss the integration of demand and supply across the supply chain and state that integration includes knowledge generation, dissemination, interpretation, and application between firms. This includes the balancing of market information and business intelligence to strategically balance demand and supply activities. Additionally, integration also includes studying the external opportunities, capabilities, and "constraints" in the supply chain from point of raw material supply to end customer demand. Therefore, it is the collection, dissemination, and shared interpretation of this busi-ness intelligence that can then allow integrated firms in the supply chain to make better strategic decisions about how to balance supply and demand. An integrated supply chain decision-making capability in itself can be an important competitive resource. Yet, it can be even more paramount when it helps them and their supply chain partners manage supply disruptions more effectively. For example, integrative decision making between firms includes approaches for managing supply support and includes identifying constraints and risks at sup-plier locations that may threaten the continuity of support to down-stream members of the supply chain and the end customer.

In the face of a supply chain disruption, such integration and deci-sion-making capabilities provide firms with the ability to quickly detect and share information about a disruption. Then having a clear view and capability to coordinate with partner firms in the supply chain, the focal firm is able to come up with a mitigation plan and more rapidly put it into action. Such decisive actions should lead to responses that are more strategically aligned among partner firms, more effective in responding to the disruption, and ultimately one that allows the recovery from the disruption to take place more quickly. However, such integrative capabilities are not always present between firms and vary greatly across the members of a supply chain network. One thing that a firm may strategically consider is how integration capabilities should be built across the network from both a vertical and horizontal perspective. For example, integration can be horizontally diverse across different functions and supply chain partners in the network, but it can also be vertically diverse from lower-level employees up to senior decision makers who can make strategic decisions. Such vertical diversity in integration may be even more important when dealing with a supply chain disruption since senior leaders are more likely to have the power and resources at their disposal to make more timely decisions and responses to help mitigate and recover from a disruption. Which raises the question, "at what levels in the firm's hierarchy should integration be accomplished across different upstream supply nodes?"

Today, firms have realized that the ability to quickly and accurately disseminate information across the supply chain to their partners about a disruption event has a significant impact on the severity of the disruption and the ability of the network to recover from the disruption event. In fact some disruptions, by their very nature, may be more difficult to detect, and this signaling ability between firms is critical to manage the occurrence of disruptions.[10] Therefore, it appears that the detection mitigation and recovery from supply chain disruptions does not rest on the shoulders of a single firm and that the ideas of collaboration, information sharing, and integration between firms may lie at the heart of creating the continuity and resiliency needed to manage supply chain disruptions.

Integration Examples

There are some excellent examples of successful companies that have used integration practices to achieve continuity in their operations and build a more resilient supply chain needed to recover from a disruption. In particular, world-class firms such as GE and Mitsubishi Electronics can provide examples of how firms have dealt with disruptions in their raw materials supply. Therefore, they serve as an example of how firms might implement practices aimed at achieving continuity and resiliency. First, we'll look at how GE manages its raw material risks and uses supply chain integration to achieve improved detection and mitigation of potential upstream supply disruptions.

GE's aircraft manufacturing and high-tech business units use large quantities of specialty metals, including rhenium, titanium, and "rare-earth" metals such as Neodymium and Lanthanum. Because the raw materials supply of these metals is critical to many of the technologies it uses in its products, GE has had to create and implement a specific risk management approach for raw materials in order to prevent disruptions and cost increases. GE's risk management approach has been lauded by industry and governments as a best practice for managing upstream supply risk for raw materials, and it includes a three-phase process that looks at (1) each particular metal's internal impact on GE, (2) the external risks that threaten the supply of that metal, and (3) the size of the business (value) that this metal creates for GE's products. As part of this process, GE has been able to identify which specific metals have a critical risk level and need to be proactively managed. For critical metals such as rhenium, which is used in aircraft engines, GE takes actions such as limiting the use of the expensive metal and increasing recycling and recovery of the metal in the downstream supply chain. More salient to our discussion, identifying raw materials with critical levels of risk allows GE to detect potential disruptions that could occur and allows the company to take mitigation actions with suppliers well in advance of an actual problem. For example, GE has built closer long-term supply relationships with existing suppliers of critical metals and has also worked with new suppliers to develop alternative sources of supply, based on its criticality analysis.[11] More recently, GE has also taken action to vertically integrate and purchase key parts of its upstream supply chain to gain even more control over supply and ensure continued access to supplies of raw materials.[12] This example at GE not only describes strategic process to detect potential supply disruptions, but it also shows how upstream supply chain integration with suppliers of critical raw materials can be a key ingredient for the management and mitigation of upstream supply risk in the network.

A second example of integration helping to manage supply chain disruptions is evident in the response to the 2011 Japanese earthquake and tsunami disaster by electronics firms, including Mitsubishi Electronics Corporation, ATMI Inc., and Arrow Electronics. Unlike the GE example, these firms were not able to fully detect or mitigate the supply chain disruption, and their experiences serve more as a lesson about the actual recovery from a major supply chain disruption. The natural disaster and related nuclear disaster in March 2011 had an immediate impact on these firms that have factories and hundreds of suppliers in the impacted region. In the aftermath of the event, Mitsubishi Electronics was able to recover from the event by limiting its production, changing to alternative suppliers, and changing product designs to avoid the use of disrupted raw materials. However, some other lessons about integration were evident during the tsunami.[13] For example, at ATMI Inc., the firm was able to analyze the impact of the event "within hours" because of a risk-mapping system that tracks information on raw materials flows in the supply chain.[14] This system was able to identify raw materials vulnerabilities two and even three tiers back in the supply chain to more quickly respond and recover from the disruption. In addition to integrating and collecting information from further back in the supply network, lessons were learned during the Japanese tsunami by Arrow Electronics about the needed level of integration and the speed needed to recover. For example, there is often a lack of a single process owner for supply chain risk, and yet during a disruption, there is a need to elevate operational-level information to the right decision makers at the right level in the firm. Since the tsunami in Japan, Arrow electronics has elevated the management of supply chain risk to a corporate level and concentrates on its decision making in the first 48 hours after an event.[15] This includes leveraging cross-functional expertise from across the firm and integration with upstream suppliers who are often dispersed across the globe. If collaborative partnerships with suppliers are already in place, it can speed the planning and flexibility needed to recover from a disruption of materials. Often, such collaboration and communication about a disruption takes place at the C-suite level in firms, where today's leaders realize that the costs of being unprepared can far outweigh the risks of having an integrated supply chain network that can respond and recover from raw material disruptions.

Integration Examples: Lessons Learned

These stories from industry should motivate you to want to dig further into the idea of supply chain integration and how it can be a capability to manage supply disruptions. One objective might be to build a framework or guiding theory to help companies in how they approach the management of their own supply risks. In doing so, a couple of key elements of such a framework already emerge from the stories we have looked at. First, successful companies such as GE, ATMI, Arrow, and others have made the commitment to invest in risk management processes to achieve visibility and integration in their upstream supply chains beyond links with their first-tier suppliers. They have identified parts of the supply network where integration and managerial attention need to be higher. Additionally, these firms have realized that integration, collaboration, and information sharing are often needed at higher levels in the organization where senior managers and C-suite executives have the resources and strategic perspective needed to respond to a disruption. Additionally, the time and speed of decision making is critical to manage supply chain disruptions and can be enhanced by such integration. Therefore, dynamic capabilities and decisive leadership can often make a huge difference in recovery time and effectiveness in managing the disruption.

Finally, these examples indicate the need to successfully allocate limited managerial resources to key nodes and regions in the supply chain network to manage supply risks for raw materials and the occurrence of supply disruptions. Additionally, achieving higher levels of integration within the firm, outside the firm, and across the supply chain network is critical, but it may also be needed at multiple management levels in that network. Therefore, we believe that firms need a network view of integration and how it can be used as an approach for managing supply risks and disruptions in a very dynamic business environment.

A Network Approach to Managing Supply Disruptions Through Integration

Many firms still approach the management of supply risks in an internal manner. For example, a single firm may be making efforts to map the flow of physical materials, parts, and products and then respond to disruptions in these flows. Current academic modeling efforts have even started to include resiliency concerns in new analytical models intended to manage risk in the supply chain.[16,17] These risk-based analytics add to the robustness of current mapping efforts by identifying weak points in the supply network and focusing managerial attention on potential high-impact disruptions to physical flows before they occur. However, there may be limits to what physical mapping and analytical modeling approaches by a single firm can achieve in managing raw materials disruption risks.

Effective supply chain integration can be an additional tool that adds to and complements analytic mapping efforts and thereby provides a more strategic and proactive capability for identifying risk and responding to a disruption. We argue that it is the relational, informational, and knowledge-based nature of integration that magnifies the ability to detect, mitigate, and recover from disruptions in a more proactive manner. When managers are caught in a paradigm of thinking about the supply network as being solely about physical structures and flows, they limit their view of the coexisting social network of relationships between the individuals in the different firms in the supply chain network. It is these individuals who communicate, collaborate, and build knowledge that allows the wider physical network to operate. Additionally, if we limit ourselves to thinking of integration as occurring only within the functions of our own firm or between two firms (buyer-supplier), then we are limiting the scope and power that integration has to offer.

Integrating beyond the firm and its first-tier suppliers offers network-level capabilities that help the firm ensure continuity of operations. It improves the ability to quickly recover from disruptions by creating transparency and gaining access and knowledge from a "broader range" of the supply network. As shown in Figure 10-1, many managers may look at collaborative and integrative capabilities as being needed only within their firm (orange) or between themselves and first-tier suppliers (blue). However, a broader network orientation and approach toward supply chain integration would allow them to see the need to strategically integrate across a variety of links and nodes in the entire supply network. For example, when risks occur at a raw material supplier multiple layers back in the supply chain (red), a firm may want to increase communication and integration back to this level. This could be especially important considering that competitors (yellow) might also be working to integrate with these same suppliers. If risks and disruption potential continue to grow at the risky supply location in the network, the firm may want to further increase the level of integration with the supplier, including building joint risk assessments and decision-making processes between the firms. This could include new contingency plans and joint investments in mitigation alternatives, such as new components and product designs, strategic inventory purchases, or creation of alternative sources of materials for the firms.

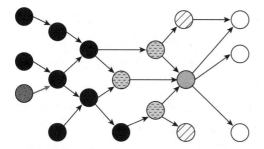

Figure 10-1 Considering Breadth of Integration in the Network

Integration across the breadth of the supply network should not be thought of as occurring solely at the operational or transactional level in the firm. Firms should not only identify the key locations in the network where integration is needed; they need to consider the actual depth of that integration at each particular network link. For example, as shown in the examples from the Japanese tsunami, firms learned that getting information to senior decision makers and C-suite leaders helped speed the decision-making processes needed to react to disruptions, and it also provided a more strategic view of the alternatives and consequences of the mitigation and recovery actions being employed. Therefore, in many cases, integration may remain at the operational level between two firms and may include only the exchange of information and lower levels of collaboration. At other times, the risks and potential for a high-impact disruption may warrant the need for more senior-level management attention and could mean that C-suite leaders from different firms work together to build new alternatives, question existing approaches, and look for creative alternatives for managing potential and actual supply disruption events.

In the end, most firms realize that there are limited resources (time, people, money) to manage risk across their entire supply networks. Adding to this complexity is that each product group in a firm may have a different supply chain with a network containing unique risks and raw materials requirements. Therefore, at the heart of managing supply chain disruptions is the idea of allocating the available managerial resources to build integration to detect, mitigate, and recover from supply disruptions at the right breadth and depth across the different supply networks. As daunting as this problem may be, it is made even more complex by the temporal and dynamic nature of our supply chains and the business world that we live in. So even if a firm makes the perfect allocation of management resources to manage supply risks across its network today, and ideally sets the right breadth and depth of that integration, it is not guaranteed to be correctly positioned for what the network risk will look like in the future. In other words, risks of supply and raw material shortages will change from week to week, month to month, and year to year in the global business world. Therefore, firms have no choice but to continue to reallocate their resources and integration efforts in the network as this risk profile changes.

Therefore, we introduce this idea of matching the level of integration to the given risk level in the network and the possible conditions that a firm may face (Figure 10-2). At any given time, the current network will have areas with different levels of disruption risk. However, how a firm allocates its management resources and integrates (breadth and depth) with other firms in different areas of the supply network may not be a match to the actual risk conditions that exist (Figure 10-2). Sometimes, we have high integration levels, where risks are low, which is a mismatch that creates wasted resources and a focus in the wrong area of the network or at the wrong level of management. Also, we may have matches in our integration levels and the existing risk in an area of the network (high integration where risk is high, and low integration where risk is low). But there is also a potential mismatch or "Gap" condition in the upper-left corner of the matrix in Figure 10-2. This is where the risk level in a part of the network is high but the level of integration is too low and does not match the need. Such a mismatch is a place in the network where a disruption is not only likely, but it is also where a disruption could really hurt the supply chain and the performance of our particular firm because integrative capabilities that ensure both continuity and resiliency are lacking. Therefore, when we have these mismatches, we must make important actions to move to one of the two other conditions, where we have a match between our integration approach and the risk level. We can do this by either increasing or decreasing the level of integration that we have with partner firms in that particular area of the supply chain network. This goes against the notion that all supply chain integration is a good thing, and it implies that there are differing and optimal levels of integration that need to be achieved at different links and nodes across the supply chain network.

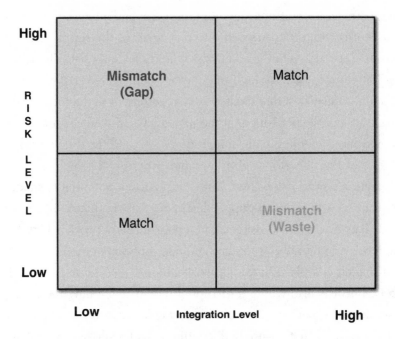

Figure 10-2 Network Risk Versus Integration Level

Because the current state of risk, especially for raw materials such as metals, commodities, water, energy, and other natural resources, is constantly changing, a firm must also periodically reanalyze whether it has a match or mismatch between its levels of integration and the current risk situation. This implies that firms must constantly reallocate their management resources and integration efforts across the network to focus on different links and nodes where risk is on the rise. This could mean that a firm may take actions such as reducing levels of operational integration in an area of the network where risks are declining, and then reallocating precious management resources to more fully integrate in areas of the network where new supply risks are emerging. This dynamic reallocation of integration capabilities and resources should also take into account both the breadth and depth of integration in order to focus senior leaders on only the most important parts of the network where integration is needed to detect, mitigate, and recover from potential disruptions.

As a firm moves from lower-level cooperative activities through collaboration to full integration with key suppliers, higher degrees of trust, resource commitment, and risk exposure are required to make that integration work. This is not always an easy task to overcome when managers try to explain to shareholders why investments in long-term risk management are impacting short-term profits, or why key proprietary information is being shared with suppliers. Therefore, firms must understand the strategic value creation and competitive advantage that can exist if they build a resilient and integrated supply network that can maintain operations when competitors are failing to operate after a disruption. Part of that value is being able to more quickly and effectively detect, mitigate, and recover from a supply chain disruption, but it could also mean taking market share from less-resilient competitors after a major disruption or ensuring the long-term survival of the firm when austere market conditions are pushing some competitors out of the market. In looking at how to manage risk across their networks, today world-class firms such as Nestle, Starbucks, GE, and others are asking tough strategic questions about who they will integrate with and at what level. For instance, Nestle is working more directly with farmers and cutting out some of the middlemen where it used to buy raw materials.[18] Similarly, other firms like Starbucks and GE are moving toward direct buys of raw materials and even ownership of raw material sources. To understand why this occurring, in the following section we take a closer look at what is happening with today's raw materials and natural resources across the globe.

Natural Resource Scarcity and the Dynamic Global Supply Network

Having a wider network view of managing supply chain disruptions through levels of integration is especially important in today's global supply chains. This can largely be explained by a number of global macroeconomic trends that are influencing and changing how and where our raw materials are supplied. Academic authors Autry et al.[19] were some of the first supply chain scholars to expand the discussion about how and why raw material supplies are being threatened in the twenty-first century. They identified how trends such as increasing global populations, emerging economies, urbanization of populations, and environmental (climate) change are making tangible changes to the way we source and supply raw materials to industrial and consumer supply chains. These economic forces are often working to create new levels of demand (and variability) in more austere locations around the globe, and suppliers may not always be able to respond to needed levels of metals, minerals, water, energy, and commodities that are often assumed to be available at a desired network location. Therefore, these new global trends have created new levels of risk several layers back in the supply network in the form of natural resource scarcity and related issues such as urban congestion and increased environmental regulations.

For example, environmental stress and climate change have influenced freshwater supplies in some congested and arid regions of the world. This water scarcity has become a major issue in regions such as Africa, India, and the U.S. Southwest where urban congestion has concentrated the number of individuals into more densely populated urban areas that may not have a sufficient water supply. Concerns about water scarcity have also resulted in increased government policy and action against firms that do not comply with water restrictions, and it has limited the ability of firms to operate at some locations. For example, declining aquifer levels in and around Beijing have drawn the attention of the Chinese government and could make it more difficult for industrial firms to move into the Beijing area if they will require a high level of water usage in their operations.[20]

In thinking about global demand for natural resources, the growing consumption of materials in China and India in the past 15 years has been nothing less than remarkable. As these economies and their populations have become more industrial and urban in nature, their consumption of raw materials has had a wide influence on the supply and prices of raw materials such as copper, iron ore, aluminum, coal, and the like around the globe. Additionally, the continued urbanization of populations in these nations has often concentrated the need for raw materials at individual locations (nodes) in the supply network, thereby increasing the risk and potential impact of a supply disruption at a critical location. Clearly these changes have not been ignored by the world economy, and rising market prices have often spurred increased exploration for minerals, material substitutions, new technologies, and new innovative ways to supply raw materials. Therefore, the message is not that global markets are at risk for an immediate wide-scale collapse, as some Malthusian scholars may warn. Instead, the message is that supply chains are constantly changing and adapting as new raw material risks emerge in the network, and supply chain managers need to take a wider view of their raw material sources in order to build specific strategies to counter new scarcity threats as they emerge.[21]

Another benefit of analyzing these wider supply networks is the realization that firms are not acting in isolation from each other and that they may often compete for limited natural resources in the same part of the network and may even interdict the supply sources of their competitors. Authors such as Markham et al. and Bell et al. have discussed how firms may actively seek (within legal limits) to delay, divert, and disrupt the raw material supplies of their competitors.[22,23] Such competition can often seem passive in nature, but can still have devastating effects if the short-term capacity of a raw material supplier is controlled by one purchaser, or if a firm builds large quantities of materials in order to make it difficult for competitors to purchase the right quantity and quality of similar materials for their own operations. When taken into account with the integration message in this chapter, this means that supply chain managers should consider integrating quickly with suppliers as risk for raw material scarcity increases in a part of the network. Otherwise, they may find that their competitors have already identified the same key supply source and built relationships and integration levels with their key supplier, and they find themselves now isolated and outmaneuvered in the raw materials network (in chess terms—checkmate!).

Some firms have additionally realized that they may need to rethink the design of their entire supply chain to create more closed-loop supply chain capabilities with increased levels of recovery and recycling that allow them to recapture valuable metals, minerals, water, and energy throughout their supply chain. This would allow them to reenter the recaptured materials back into the forward supply chain, thereby replacing the need to purchase more virgin materials from raw material suppliers.[24] These high-level capabilities move beyond the firm level and may require higher levels of integration with raw materials suppliers (and their engineers) who can help build the processes and operations needed in a closed loop supply chain. Such a redesign may take collaborations and partnerships between firms and require them to make significant investment in new technologies and materials innovations that can help them manage raw materials risks and create competitive advantages for each firm.

When we manage our supply chains, we often think about nonrenewable natural resources, such as the oil, metals, and minerals used to build high-tech products, such as computers, automobiles, and airplanes that define our daily lives in this century. But more important may be some of the renewable natural resource supply chains, such as agricultural crops, timber, and fisheries, that are significant sources of raw materials for our consumer and food supply chains. In 2014, the fisheries supply chain was highlighted by research at the University of British Columbia as being a vulnerable supply chain, fraught with illegal activity and lack of transparency.[25] This research, which caught the attention of the White House and U.S. regulators, contends that up to 32 percent of the fish sold in the U.S. has been caught illegally in the oceans of the world. Since these regulations were put in place to ensure the "renewable" nature of the fisheries, it is clear that risks are increasing in the raw material supply of fish in our food supply chains, and it is believed that as much as 80 percent of the world's fisheries are already being overfished.[26]

Yet information about the actual source of supply for fish in the United States is very poor, and currently there is a real lack of transparency of the makeup of the network and where risk is originating in the network. It is clear that U.S. suppliers and retailers need to increase their integration with upstream wholesalers, canneries, and even the fishermen themselves to ensure integrity in our fishery supply chains. Additionally, supply chain integration could provide critical abilities to detect and mitigate potential disruptions related to perishable fish products that could harm consumers through foodborne illnesses. Obviously, the impact of such a disruption could be very high and the potential cost to the company tremendous. However, the dynamic and global nature of fishing fleets and the network that supplies fresh fish to our supply chains will not make such integration easy. It is quite possible that new technologies will be needed to communicate and collaborate with fishing fleets, canneries, transportation providers, and suppliers in order to monitor fishery risks and ensure much-needed levels of continuity and resiliency in this vulnerable raw material supply chain.

Conclusion

Supply chain integration can be a strategic capability that helps to manage disruptions in upstream flow of raw materials and one that builds both continuity and resiliency for the firm. However, a network view of the supply chain is needed to determine where limited management resources should be committed to create integration and management of such risks. This includes mapping and understanding the breadth of the network itself and also determining the right managerial depth of integration. However, the changing global environment means dynamic changes to our natural resource supply sources and the potential for new and emerging disruptions in our upstream raw material suppliers that may be several levels upstream in the supply chain. In fact, some supply chains, such as fisheries, are truly vulnerable and are badly in need of integration capabilities to shore up (no pun intended) these risks and rebuild resiliency. Therefore, we have offered a framework and approach that managers can use to strategically allocate their own resources (time, people, money) to integrate at the right locations in the network and thereby match the level of integration to the risk at a specific region of the supply network. This implies that more or less integration may be needed at different areas in the network to efficiently ensure continuity and resiliency in the supply chain. However, managers need to also consider the dynamics of managing risk and disruptions in the upstream supply network, and they must continuously reallocate and change their integration approach in different parts of the network to realign their management resources toward the most risky parts of their networks. In the end, supply chain integration can be a dynamic capability that helps the firm overcome resource scarcity and supply disruptions in its upstream supply chain. Such a capability can help ensure the long-term performance and competitive advantage of firms in the supply chains of today and tomorrow.

Endnotes

1. John E. Bell is an Associate Professor of Supply Chain Management and Christopher W. Craighead is the Dove Professor of Supply Chain Management, both at the University of Tennessee's Haslam College of Business.

2. O'Marah, K. 2015. Coca-Cola and Water: The Pause That Refreshes. SCM World, Beyond Supply Chain Weekly Updates, August 28. http://www.scmworld.com.

3. Hignett, A. 2015. Cadbury Egg-Maker Sees Profits Melting. February 12. http://www.cityam.com.

4. Plimmer, G. 2015. Global Supply Chains Face Heightened Risks. *Financial Times*, May 10. http://www.ft.com.

5. Lin, K. and E. Chasan. 2015. In Conflict Minerals, Ethical Investors Gain Ability to Rank Companies. *Wall Street Journal*, August 5. http://www.wsj.com.

6. Lambert, D. and M. Cooper. 2000. Issues in Supply Chain Management. *Industrial Marketing Management* 29: 65–83.

7. Blackhurst, J., K. Dunn, and C.W. Craighead. 2011. An Empirically-Derived Framework of Global Supply Resiliency, *Journal of Business Logistics* (32)4: 374–391.

8. Wang, Q., C. Craighead, and J.J. Li. 2014. Justice Served: Mitigating Damaged Trust Stemming from Supply Chain Disruptions. *Journal of Operations Management* 32: 374–386.

9. Esper, T., A. Ellinger, T. Stank, D. Flint, and M. Moon. 2010. Demand and Supply Integration: A Conceptual Framework of Value Creation Through Knowledge Management. *Journal of the Academy of Marketing Sciences* 38: 5–18.

10. Kleindorfer and Saad. 2005. Managing Disruption Risks in Supply Chains. *Production and Operations Management* 14(1): 53–68.

11. Duclos, S., J. Otto, and D. G. Konitzer. 2010. Design in an Era of Constrained Resources. *Mechanical Engineering*, September: 36–40.

12. Linebaugh, K. 2013. GE Brings Engine Work Back. *Wall Street Journal*, February 6. www.wsj.com.

13. Logistics Institute–Asia Europe. 2011. Combating Supply Chain Disruptions: Lessons Learned from Japan. *Think Executive*, November.

14. Driscoll M. 2013. Supply Chain Disruption Risks: Ask the Right Questions. *Financial Executive*, October. http://www.financialexecutives.org.

15. Ibid.

16. Harrison, T.P., P.J. Houm, D.T. Thomas, and C.W. Craighead. 2013. Supply Chain Disruptions Are Inevitable—Get READI: Readiness Enhancement Analysis via Deletion and Insertion. *Transportation Journal* 52(2): 264–276.

17. Deane, J.K., C.W. Craighead, and C.T. Ragsdale. 2009. Mitigating Environmental and Density Risk in Global Sourcing. *International Journal of Physical Distribution & Logistics Management* 39: 861–883.

18. Nestle to Double Its Commodity Food Suppliers. 2014. Food Logistics, June 30. http://www.foodlogistics.com.

19. Autry, C., T. Goldsby, and J. Bell. 2012. *Global MacroTrends and Their Impact on Supply Chain Management*. New York: Pearson Financial Times.

20. Cui, X., G. Huang, W. Chen, and A. Morse. 2009. Threatening of Water Change on Water Resources and Supply: Case Study of North China. *Desalination* 248: 476–478.

21. Bell, J.E., C.W. Autry, D.A. Mollenkopf, and L.M. Thornton. 2012. A Natural Resource Scarcity Typology: Theoretical Foundations and Strategic Implications for Supply Chain Management. *Journal of Business Logistics* 33(2): 158–166.

22. Markman, G., P.T. Gianiodis, and A.K. Buchholtz. 2009. Factor Market Rivalry. *Academy of Management Review* 34(3): 423–441.

23. Bell, J.E., C.W. Autry, and S.E. Griffis. "Supply Chain Interdiction as a Competitive Weapon. *Transportation Journal* 54(1): 89–103.

24. Bell, J.E., D.A. Mollenkopf, and H.J. Stoltze. 2013. Natural Resource Scarcity and the Closed-Loop Supply Chain: A Resource Advantage View. *International Journal of Physical Distribution & Logistics Management* 43(5/6): 351–379.

25. Pramod, G., K. Nakamura, T. Pitcher, and L. Delagran. 2014. Estimates of Illegal and Unreported Fish in Seafood Imports to the US. *Marine Policy* 48: 102–113.

26. Fears, Darryl. 2014. Illegally Caught Fish a Big Part of US Imports. *Washington Post*, April 21; reprinted online by *Boston Globe*, May 2. http://www.bostonglobe.com.

11

Integrating Ideas and Environments: Blending Marketing Strategy with Context for Organizational Success

By Kelly Hewett, Adam Hepworth, and Sharon Watson[1]

Faced with turbulent markets characterized by slow economic conditions and competitive pressures, many firms are looking to implement innovative marketing strategies. Concurrently, marketers are increasingly focused on demonstrating their contribution to the firm's bottom line, and innovation is one avenue for doing so. Innovation associated strictly with product improvements is a strategy that, while still valued, may not always lead to sustainable competitive advantages in the way product innovation strategy once had. Instead, the rapid advancements of technology have paved the way for quicker and less-costly product innovations and opened the door to participants competing based on product innovation. Firms now recognize that innovation is not limited to product development, and greater attention is being paid to marketing strategy innovation, or the production or emergence of new ideas.

Coca-Cola recently leveraged marketing strategy innovation when the company began the formidable initiative of rebranding Coke and turning around slumping sales that had plagued the industry for nearly a decade. The *Share a Coke* marketing campaign initially began as

Project Connect in Australia in 2011. After executives introduced the campaign stateside, it was widely touted as the most impactful and successful marketing campaign for the summer of 2014. The ingenious idea of personalizing Coke bottles and cans with eponymous labels led to a resurgence in unit sales and bolstered Coke's revenue. Far more critical, the uptick in sales was enough to derail the decade-long stint of declining sales and reverse the downward trend that many of Coke's direct competitors, notably Pepsi and Dr. Pepper, are currently still experiencing. Coke's reversal of fortune and climbing soft-drink sales proved enough in extending and expanding the Share a Coke marketing initiative from a short-lived fad into a long-term concentrated effort of branding identity. Up from 250 of the most popular names featured on Coke cans, the company quadrupled its repository of names, committing to preserving its socially targeted campaign.[2]

As was the case with Coca-Cola's Share a Coke campaign, innovation and creativity go hand-in-hand, yet the two concepts are distinct. Creativity and innovation together constitute an overall process aimed at introducing new ways of doing things. The creativity stage focuses primarily on idea generation, whereas the innovation stage includes the implementation of those ideas in an effort to enhance processes, organizational practices, or products and services. This view of creativity at earlier stages and innovation at latter stages of the overall process is echoed across disciplines in organizational psychology, management, and innovation.

The relatively sparse academic research on marketing innovation focuses largely on performance outcomes in the latter stages and on the drivers of marketing program and manager creativity.[3] Rather than focus on marketing programs, which encompass decisions relating to pricing, promotions, and distribution, the study reported here emphasizes innovation in the more comprehensive perspective of marketing strategy. Marketing strategy encompasses broader activities, such as situation analysis, developing and managing marketing assets and capabilities, strategies with regard to managing relationships with stakeholders, and resource allocation decisions, as opposed to decisions regarding strictly a product's marketing programs. An example of a company leveraging its marketing assets and capabilities as well as its stakeholder relationships is Procter & Gamble's shift to "win-win partnerships" in 2000. Historically, P&G had focused on short-term transactional sales negotiations with customers designed to incrementally and temporally increase sales performance. Counter to short-term sales objectives, win-win partnerships are based on collaboration and keeping the best interests of the partner at the forefront of considerations. P&G capitalized on such partnerships to pioneer a new form of relationship concentrated on creating and sustaining mutual value for both P&G and its customers. In turn, the shift to a long-term-oriented relationship provided P&G with a reliable source of income from enhanced customer retention efforts.[4]

As in the case of P&G, the focus for the study at hand is specifically on the extent to which organizations emphasize innovation in marketing strategy. In general, promoting innovation is considered important for firm performance. Academic research has demonstrated that firms experience enhanced organizational performance in part due to an emphasis on innovation. However, innovative ideas are frequently deemed risky, so they are prone to delays or often not implemented at all. As noted by researchers Dada and Watson in their study of entrepreneurship in franchising, a departure from proven procedures can be viewed as posing risk to the overall firm and therefore may not necessarily be supported.[5] Such departures may draw opposition from risk-averse individuals throughout the firm and stymie attempts to innovate.

The results of the research study reported here demonstrate that marketing strategy innovation has the greatest impact when integrated appropriately with a firm's internal and external environmental conditions. In particular, the study proposes and finds evidence to support the notion that firms can enhance performance with better integration between marketing innovation and internal conditions such as firm structure and the work environment, and external market conditions. We define integration as instilling a unity of effort across organizational levels and cross-functional subunits and promoting fit with the organization's external environment, and then aligning those efforts with the firm's strategic focus. With respect to internal considerations, the study results indicate that marketing innovation is most effective when firms give managers more autonomy in their decision making. Managers repeatedly described excessive centralization as hampering innovation and autonomy as enabling innovative strategies. On the other hand, the study offers evidence that innovation is less appropriate with centralized structures, suggesting that the level of marketing strategy innovation needs to be integrated or aligned with firm structural characteristics. Similarly, with respect to the firm's external environment, marketing strategy innovation may not yield the same benefits in stable as opposed to turbulent environments. As such, this study addresses the overarching question: Under what set of circumstances is marketing strategy innovation likely to be most effective? Also, under what combination of conditions would marketing strategy innovation not be effective?

When marketing strategy innovation fails to consider internal and external considerations, companies rarely walk away unscathed. Recently, Amazon executives launched a frenzy of advertisements designed to spur online purchases for its Amazon Prime Day event. Although Prime Day may have ignited single-day online sales for the e-retailer, it also unleashed indignation from furious Prime members who complained about lackluster deals and underwhelming product selection. It is therefore evident that optimal strategy innovation must be aligned with the firm's internal or external environment or else it risks rendering the innovation strategy obsolete, or as the Amazon example demonstrates, alienating a firm's loyal customer base.

This study proposes and tests a model linking marketing strategy innovation and conditions under which this approach is likely to be most effective. We view marketing strategy innovation as a key strategic variable that, when aligned accordingly with the firm's internal and external environment, leads to profitable outcomes. From the academic literature, configuration theory postulates that an ideal set of internal firm conditions can fit with a firm's external context to improve performance.[6] The greater the extent to which a firm's circumstances are similar to this ideal set of conditions, the greater its "fit," or integration with that profile is considered to be. Marketing scholars have a long tradition of examining how characteristics of the environment influence various aspects of sales management and global marketing strategies along the standardization-adaptation continuum. Researchers in the organizational behavior field have examined similar influences on creativity, and the study reported here is unique in its exploration of fit between a firm's environment and its emphasis on marketing strategy innovation.

Based on evidence from previous research in the strategy and marketing fields, we propose that critical internal conditions—namely, an autonomous yet challenging work environment—when integrated with marketing strategy innovation, can positively influence performance. In addition, a firm that is able to provide an environment in which managers have access to abundant knowledge and insights regarding the marketing environment is essential. Last, we propose that a turbulent external environment, in which tastes and preferences of buyers as well as the state of competition is changing rapidly, provides opportunity for marketing strategy innovation and is therefore more likely to generate positive performance benefits. Figure 11-1 provides a graphic depiction of the concept of integration between marketing strategy innovation and a firm's internal and external conditions.

Figure 11-1 Integration of Marketing Strategy Innovation with Firm Conditions

We employ a multimethod approach using survey data gathered from senior marketing managers as well as in-depth interviews to triangulate the findings from the survey. The results support our contention that marketing strategy innovation can lead to enhanced performance under certain internal and external conditions. Also highlighted in the study are situations in which emphasizing marketing strategy innovation does not ultimately enhance performance. It could be argued that marketing managers may be more likely to assume innovative marketing strategies are more effective; therefore, they may be more likely to face the question of how to encourage innovation, rather than consider whether encouraging innovation is the best course of action for the firm. This study offers evidence that innovative strategies are not necessarily readily embraced by all individuals in the firm, nor are they necessarily effective under all conditions. Thus, a key takeaway from the study reported here is an integrative assessment of internal and external firm factors that, when integrated with marketing innovation encouragement, will be associated with better performance. The results of our study also help shed light on how and when marketing strategy innovation should be integrated with a firm's internal and external conditions. That is, we highlight the specific conditions innovative firms should look for in attempting to build and implement innovative marketing strategies.

The Study: Integrating Marketing Strategy with a Firm's Internal and External Conditions

The study consisted of two parts: a mail survey and a set of in-depth interviews with marketing managers.

The Survey Study

Given the focus on marketing strategy innovation, the goal in conducting our survey was to identify firms where marketing was a critical function in terms of how they compete. As a result, advertising expenditure was used as the sample selection criterion. The targeted sample consisted of 1,209 U.S. business units with advertising expenditures of at least $1.5 million. We addressed the surveys to a senior marketing executive at each firm. We selected marketing managers at this level as key informants because they are assumed to be the most knowledgeable about questions regarding marketing strategy, which was the focus of the survey. A total of 195 usable responses was received, for a response rate of 17.18 percent, which is typical for this type of survey among senior managers.

When completing the survey, managers were instructed to focus on the innovation of their recent marketing strategies. To measure the key variable, *marketing strategy innovation*, the survey asked questions such as "In this organization, marketing employees are encouraged to seek innovative strategies to compete in the marketplace," and "In this organization, marketing employees who suggest new and different strategies are well supported by senior management." Respondents rated their agreement with a series of such statements.

Given that the purpose of the study was to understand how marketing strategy innovation should be integrated with other characteristics of the firm, the survey asked about three important aspects of the firm's internal conditions: the level of *autonomy*, the *challenge* presented by the work environment, and *knowledge* of the marketing environment. With respect to the firm's level of autonomy, we were interested in the degree of freedom given to marketing employees with respect to decision making, which we measured by asking respondents to rate their agreement with statements including "In this organization, marketing employees have sufficient freedom in developing strategy," and "In this organization, marketing employees have flexibility in getting work done."

Additionally, we sought to understand how challenging the work was for personnel in the marketing area. The survey assessed challenging work by asking respondents to rate their agreement with statements about the marketing function, such as "Marketing jobs in this organization are intellectually demanding," and "Marketing jobs in this organization involve frequently responding to difficult challenges in the marketplace." Questions used to assess knowledge of the marketing environment asked about the organization's knowledge of specific elements of marketing, such as channel member behavior; customer motivation; customer purchase and usage behaviors; and relevant political, legal, technological, and demographic trends.

It is important that the level of marketing strategy innovation be integrated not only with firm conditions, but also with the external conditions faced by the company. With respect to external conditions, the aspect of the marketing environment that is most relevant to marketing strategy innovation is the degree to which the external environment, including factors such as customers and competitors, is changing. Because marketing innovation is crucial in rapidly changing

markets, we asked respondents to rate their agreement with statements regarding *market turbulence*, such as "In our business, customers' product preferences change substantially over time," and "New
customers have product-related needs that are different from those of
our existing customers."

Finally, to better understand when integration of marketing
strategy innovation is warranted and when it is not, the survey also
included questions aimed at assessing firm performance. Because firm
performance is multidimensional, we asked about three aspects of
performance: profitability (return on assets), market share, and sales.
We asked respondents to evaluate each of these three performance
dimensions compared to their competitors, as well as compared to
their internal objectives. Thus, there was a total of six questions that
measured performance.

The Interviews

In-depth interviews with senior marketing managers from a variety of industries were used to triangulate the survey findings and generate further understanding of the underlying phenomena. Managers
in industries similar to the survey participants' (services, industrial
goods, high technology, and consumer products) were contacted in an
attempt to mirror the results of the survey. In all, 42 interviews were
conducted (14 face-to-face at respondents' offices; 28 via telephone).
All participants held managerial positions and were involved in developing marketing strategy; none overlapped with the survey sample.
Interviews probed the essence of innovative marketing strategy as
well as work environment features. To encourage candid responses
during the interviews, which lasted 30 to 90 minutes, respondents
were promised anonymity. The information gleaned during these
interviews enhances the information gathered through the survey,
so the results of both the interviews and the survey are discussed
together in the sections that follow.

Results

The results of the survey study suggest that firms can, in fact, enhance performance the greater the integration between marketing innovation encouragement and internal and external conditions. Relationships were tested in terms of multiple conditions versus traditional approaches with interactions of marketing strategy that examine the conditions independently. Each internal condition—autonomy, challenging work, and market environment knowledge—along with the external condition of market turbulence was individually evaluated to examine the effect each condition ultimately had on organizational performance.

Effect of Autonomy

Autonomy was characterized as the level of freedom given to marketing managers, especially with respect to decision making. The key issue examined here is the fit between autonomy and marketing strategy innovation. Results did, indeed, indicate a clear link between autonomy and marketing innovation. Marketing managers repeatedly described excessive centralization of firm operations and governance structure as an impediment to innovation, whereas an autonomous internal environment promoted innovative strategies. In one such response, autonomy was deemed beneficial because "it doesn't have to go up this huge chain of command for an innovative idea to be tried." Employees in this respondent's particular firm considered autonomy as a form of individual empowerment that contributes to the success of these strategies. A manager stated the innovativeness of his firm's internal structure is characterized by flexibility and scenario-based choices: "We give staff enough leeway to evaluate situations and escalate if/as necessary. We even run them through scenarios and evaluate their choices." Using scenarios created a sense of freedom to develop innovative strategies and led to positive performance measures.

In contrast to the flexible internal environment of autonomous firms, evidence was found suggesting marketing strategy innovation is less appropriate for firms with centralized structures. One manager indicated that the extent to which his employees can innovate is limited, and that his firm encourages a focus on areas where they "can make an impact," which occurred through managerially controlled facilitation intended to guide employees' actions. Such an emphasis on control and formal internal structure was believed to create stronger organizational performance. Another manager described his firm as excelling due to an ability to be "profitable and efficient in our expense profile compared to competitors," and suggested a key factor was a "bureaucratic approach" in which the marketing team generally uses "what has worked before because it has always worked for us." He also reflected on a period in which "a lack of discipline became chaos." In instances in which managers exhibited more control, formalized structure and avoidance of unpredictable outcomes associated with innovation guided the actions of management, leading them to implement strategies that mirrored past successes. In such firms, less of an emphasis on marketing strategy innovation may enable closer control of employee efforts.

Effect of Challenging Work

In addition to establishing an internal environment favorable to autonomy, firms emphasizing marketing strategy innovation were typified by challenging work environments. Challenging workloads can stimulate worker productivity and elicit personal achievement, as employees are motivated by the challenge associated with the task at hand.[7] Moreover, challenging work has been shown to increase employee job satisfaction, which leads to several other positive outcomes related to worker productivity and overall employee well-being. According to one manager, "work enrichment" efforts to provide a more challenging internal environment enabled marketing strategy innovation. Another respondent observed that if managers are "challenged to think of new ideas, new ways to develop business, it broadens our scope, it fosters innovation for sure. If you don't have a challenging environment, and at times a competitive environment, then you may get a little complacent in what you are doing." Commenting further on the need for challenging work, one manager mentioned that challenge in professional roles "keeps the job interesting and raises the need for creativity" by raising "more questions, obstacles and the need to think about how to do things better." Work that is not challenging, on the other hand, gives rise to "boredom and complacency." Considerable research into the employee work environment has provided evidence that workplace challenges motivate employees to rise to the occasion and result in a sense of personal achievement as well as worker development.[8]

However, challenging work is not feasible in every environment, nor is making work more challenging always appropriate in terms of supporting innovative marketing strategies. One manager dismissed the idea of innovative marketing strategies as being critical to a firm's success, emphasizing that his firm's unit leads in market share and the internal atmosphere of the firm ensures managers "already have enough to do." Another manager echoed similar sentiments by stating that his firm's traditional marketing approach remained "more reactive than proactive." However, overall strong firm performance persisted despite not using innovative marketing strategies. Regarding creativity in marketing, one manager advised that challenging work that encourages creative marketing endeavors also includes risks for employees: "Like any organization, they all say that they encourage risk-taking but then sacrifice those that do." Therefore, employees run the risk of having management perceive their marketing innovation initiatives as too far-reaching and, therefore, may risk their job security. Although challenging work was linked to marketing strategy innovation, neither the challenging environment nor innovation was critical in all situations. Rather, the integration between challenging work and marketing innovation was more important.

Knowledge of the Marketing Environment

Knowledge of the firm's marketing environment was also revealed to be an essential internal characteristic for firms pursuing marketing strategy innovation. As elicited in interviews with marketing managers across industries, such knowledge was predominantly described as resulting from close customer relationships and as enabling firms to develop innovative strategies to meet those needs. In one such instance, a manager acknowledged that marketing strategy innovation had increased after his firm restructured around its customers to foster understanding of needs. Marketing performance was reduced by atrophied market learning, as his firm fixated on existing demand and ignored market changes. Another marketing manager saw his firm as a marketing "innovator" and credited his firm's marketing research with allowing it to "stay ahead of trends and innovations in the marketplace." Emphasizing marketing strategy innovation was seen as enabling a superior market position for his firm: "Marketing innovation expands the number of customers we can touch...it expands our image and our brand." A marketing manager from an industrial business-to-business company offered his perspective on the importance of linking creativity with knowledge of the marketing environment to stay relevant to clients:

> Not every idea is going to be great, I mean you are going to evaluate that, but if you don't consistently evaluate new ideas and look at new ways of doing things all you're doing is executing the same thing over and over again, and while you may be really good at doing that and you can extend your life for a while, in the end I don't think that you are going to be a really viable company.

Not all managers judged such knowledge to be an essential precursor to organizational performance. Specifically, many of our interview respondents commented on how focusing tightly on learning and marketing strategy innovation might not always be critical for performance. In fact, one marketing manager described her firm as relatively unconcerned with learning about its environment. She candidly stated that the employees are primarily "worried about the bottom line, and they know that they have to use extra resources and spend extra money to obtain information...basically you have to attain your objectives." She attributed a focus on the bottom line with the reason why her firm is able to maintain a dominant market position. This firm's example provides insight into how a focus on learning and marketing strategy innovation is not the exclusive route to enhanced performance. Instead, the integration between knowledge of the marketing environment and marketing strategy innovation was a more appropriate course than focusing on either alone.

Within the past decade, few companies have leveraged fit between knowledge of the marketing environment and marketing innovation better than HubSpot. The Boston-area company carved its niche in the online marketing community by revolutionizing how companies drive traffic to their websites. Instead of utilizing traditional marketing avenues of advertising and promotion, HubSpot advises its clients to unleash a calculated strategy of blogs, educational papers, social publishing, and SEO in an effort to reach potential customers with similar interests. In doing so, the strategic release of content drives consumers to client websites and results in higher lead conversions. HubSpot works closely with its clients in renovating their existing and often outdated marketing strategies to reach new customers through an innovative strategy called inbound marketing—bringing visitors to the company. It was a thorough understanding of the changing marketing environment (the desire for closer client relationships) coupled with shifts in consumer preferences for timely and relevant content that allowed HubSpot to leverage its innovative strategy of inbound marketing. Abandoning traditional approaches of hounding potential clients with cold calls and impersonal emails or exhausting substantial advertising expenses has, in turn, given HubSpot a sustainable competitive advantage.[9]

Additionally, Coca-Cola's resurgence in the past year is directly attributable to the company's knowledge of its marketing environment. The brand's iconic image connected with older generations but appealed far less to the younger generation's interests. After realizing Coca-Cola no longer resonated with teenagers and millennials, the soft-drink juggernaut needed a way to connect with its younger demographic. Creative marketing efforts such as the Share a Coke campaign described earlier would have been misdirected and costly without the essential insights gathered from research into Coke's younger customers. Such an investment in market research aligned Coke's message with the interests of the teenager and millennial population and led to an outright turnaround in overall sales.

Market Turbulence

In addition to internal firm characteristics being integrated with marketing innovation, our analysis of the interviews also revealed that firms operating in turbulent markets realized stronger performance records when emphasizing marketing strategy innovation. One manager described the "volatility of changing needs and changing preferences of customers" in his market and suggested that his firm's innovative focus led to brand leadership. Another described his market as "dynamic," and his firm as needing to "adapt to change" and be "flexible" by focusing on innovative marketing strategies. As mentioned previously, few companies have orchestrated a more comprehensive rebranding effort—and none certainly more complex and worldwide—as Coke in adapting to evolving customer preferences. Coke's innovative marketing strategy represented an unprecedented level of global mass personalization for a product. Changing customer needs were seen as driving a focus on innovation.

Alternatively, the failure of Amazon to accurately assess market conditions and understand its customers' preferences resulted in irate Amazon Prime customers during its inaugural Amazon Prime Day event. With deals promising to rival Black Friday sales, the innovative mid-year sales frenzy did accomplish the firm's bottom-line objective of increasing sales revenue; however, it resulted in a backlash of irritated Prime members expressing their displeasure over the quality of deals.[10] Many complained that Amazon essentially staged an online garage sale of Tupperware and other unexciting products that fell short of the expectations of their highest-valued customers. While Prime members are bargain hunters, they still seek value and quality in online shopping. Innovative marketing strategy alone is not sufficient for ideal firm performance. Instead, as this example illustrates, integration of market conditions and marketing innovation is more imperative than strictly encouraging marketing strategy innovation under any conditions.

However, such innovative marketing strategies may not always be necessary, or even appropriate, in every industry or environment. One interview respondent indicated that his firm's customer base is stable, with "clearly defined needs." He attributed his firm's success to "discipline of the management team." Another respondent who described her firm's market as stable described a tendency for her firm to be more "incremental" in developing marketing strategy, indicating that their success has come from building on what has already been done. Under conditions of stable customer preferences, there may be less of a need for developing innovative marketing strategies and risking alienating your customers.

Customer alienation and infuriation also evidently resulted from Apple's release of U2's thirteenth album *Songs of Innocence* when the album automatically downloaded to iTunes users' accounts everywhere. Apple CEO Tim Cook originally predicted that the album launch would be "the largest album release of all time."[11] However, in an industry in which music preferences remain relatively stable, yet vary considerably from person to person, many viewed the free album as intrusive and an unwelcomed download. Although the album's promotion and launch represented a momentous and innovative marketing strategy, the event was poorly received by innumerable iTunes users. Apple eventually weathered the onslaught of negative attention that played out on multiple social media platforms, but not before having to construct an uninstallation web page on its website for users to remove the U2 album. It can be logically concluded that Apple's marketing strategy innovation erroneously failed to consider the music preferences of its iTunes users under conditions of a relatively stable market.

Overall Organizational Integration

Thus far, we have investigated how internal firm characteristics and external market conditions should be integrated with marketing innovation. However, these conditions are not completely independent, nor can firms address them individually. Rather, better performance typically is achieved by aligning the level of marketing innovation with the *set* of conditions, or system, in which the firm operates. Thus, in addition to examining how marketing innovation should be integrated with individual internal firm and external market conditions, we also assessed whether matching the level of marketing strategy innovation to the overall set of conditions facing the firm would lead to better performance. In essence, we sought to understand if there was an ideal set of conditions, both internal and external, in which marketing innovation would have the most impact on firm performance, and the set of conditions in which marketing innovation is not warranted.

To investigate this question, we needed to identify the set of internal and external conditions associated with high-performance firms with high levels of marketing strategy innovation, as well as the set of conditions associated with similarly high-performing firms that had low levels of marketing strategy innovation. We characterized high-performance firms as those that ranked in the top 10 percent on our performance measure. We segmented these firms into high marketing strategy innovators and low marketing strategy innovators, based on their survey responses. This procedure gave us two

groups of high-performing firms: one group that was high on marketing strategy innovation and the other group that was low in marketing
strategy innovation. We then created "ideal profiles" for each group
by calculating the mean value for each of the internal and external
conditions. Table 11-1 indicates the mean values of each condition
for high-performing firms in both the high and low marketing strategy
innovation groups. These ideal profiles indicate how best to integrate
the level of marketing strategy innovation with the set of conditions
the firm faces.

Table 11-1 Ideal Profile Analysis Results

	Top-Performing Firms with High Marketing Strategy Innovation	Top-Performing Firms with Low Marketing Strategy Innovation
Challenging Work	5.58	5.48
Autonomy	5.48	3.96
Knowledge of Marketing Environment	5.23	4.44
Market Turbulence	4.15	4.93

The results of the profile analysis present interesting differences
between the two sets of high-performing firms on several internal
and external conditions. One of the most important findings is that
in the high-performing firms with high levels of marketing strategy
innovation, marketing managers are working with a great deal of
autonomy. In stark contrast, the firms with low levels of marketing
strategy innovation that are high performers report much lower levels of autonomy. This result suggests that there is a clear fit between
job autonomy and marketing strategy innovation that leads to higher
performance. Indeed, it suggests that marketing strategy innovation
is most effective when marketing managers have greater freedom
in their decision making in the implementation of these innovative
strategies. This finding also indicates that in firms in which marketing
strategy is less innovative, higher levels of autonomy do not lead to

better performance, but rather, less autonomy is a better fit in such situations.

Additionally, marketing managers with better knowledge of the marketing environment were a better fit with high levels of marketing strategy innovation, whereas in firms with lower levels of marketing strategy innovation, less knowledge of the marketing environment was necessary to achieve high performance. In other words, marketing managers tasked with developing and implementing innovative strategies had much more success if they had not only a good deal of freedom in their decision making, but also a high level of knowledge about market conditions. These two characteristics, autonomy and knowledge of the market environment, go hand-in-hand, for it is only those marketing managers with high levels of knowledge of their markets that will be able to make effective marketing decisions when given the autonomy to do so. Our findings also indicate that lower autonomy, coupled with a somewhat lower level of market knowledge, can be effective as well, but more so in firms that are emphasizing a less-innovative marketing strategy.

One finding regarding the ideal profiles that runs contrary to what one might expect, and at least in part with the input from our interview participants, is with respect to external market turbulence. We found that the higher-performing firms in the low marketing strategy innovation group faced more turbulent markets than the set of high-performing firms with more innovative strategies. It can be argued that innovative strategies are necessary when customer needs are rapidly changing, but in our survey study we find that high-performing firms with less innovative strategies face more turbulent markets than their more innovative counterparts. One explanation for this unusual finding could be that when customer needs are changing too rapidly, it is difficult to predict the direction in which marketing strategy innovation should be directed. Hence, a more traditional strategy may be needed until the emerging customer needs are better understood.

To test whether these profiles are, indeed, ideal for high and low levels of marketing innovation, we then tested whether a firm's deviation from the ideal profile would result in lower performance. To conduct this analysis, we first divided the sample of firms into low and high innovators, based on their scores on the marketing strategy innovation measure. We then calculated the Euclidean distance of each firm from its ideal profile using the four dimensions in Table 11-1. In other words, for a firm that was a high marketing innovator, we calculated its multidimensional distance from the ideal profile of top-performing firms in the first column of Table 11-1. We did the same for all firms high in marketing innovation. For firms that were low in marketing innovation, we calculated their distances from the ideal profile in the second column of Table 11-1. We then correlated the firms' deviations from their ideal profiles with firm performance and found a statistically significant negative relationship. This analysis demonstrates that the greater a firm's set of conditions deviated from the ideal profile set of conditions for high or low levels of marketing innovation, the lower performance tended to be. The significant negative correlation between the distance from the ideal profile and performance supports the idea that a firm's level of marketing innovation should be integrated with the overall set of conditions facing that particular firm. In other words, under one integrated set of conditions, high levels of marketing strategy innovation lead to better performance, whereas a low level of marketing strategy innovation is better for firm performance under another integrated set of conditions.

Implications

Findings from the study indicate that the organization of marketing activities can, indeed, enable strategy implementation. Specifically, the results demonstrate the value of taking into consideration a firm's internal characteristics and external environment when implementing marketing strategy innovation, and underscore the relevance of a systems view in assessing the value of a strategy-conditions integration. The results corroborate a view that a focus on encouraging marketing innovation may not always be the most effective approach to achieving market advantages in all situations, despite considerable attention to this issue in prior studies. For instance, under conditions of low market turbulence, the findings suggest that firms should place less of an emphasis on innovation and direct resources of the firm to other activities, instead of innovation. The findings support the notion that managers should not necessarily readily encourage innovative marketing strategies; rather, they should assume boundary-spanning roles and consider the conditions under which such strategies may be most effective.

Recent research supports our findings that aligning marketing innovation with internal characteristics and the external environment of the firm impacts the overall success of such innovation initiatives. A study by Dibrell, Fairclough, and Davis reports that the coordination of internal and external processes to match the pace of change in competitive and technological cycles influences a firm's ability to innovate, echoing the importance of integrating internal and external conditions.[12] Furthermore, their study confirms that focusing exclusively on internal characteristics limits the firm's ability to innovate when compared with a strategy that dually stresses internal processes coupled with the external environment.

The study reported here also provides managers with insight into the extent to which they should focus on marketing strategy innovation. The results imply that managers may be better able to enhance performance by determining whether marketing strategy innovation should be encouraged based on the conditions they face internally. Although even internal conditions of autonomy and knowledge of the marketing environment may not be immediately augmented at the managerial level, understanding the internal conditions optimal for marketing strategy innovation is crucial for long-term focus on innovation and allows managers to align strategy with those conditions. Having a solid understanding of how the firm interacts with internal and external conditions beyond its control is a necessary precursor to properly assessing the appropriate course of action relative to marketing strategy innovation. Whatever situation may arise, whether anticipated or unforeseen, certain internal and external characteristics play a major role in determining the overall success of marketing strategy innovation. In general, it is critical for managers to understand how strategy interacts with the environment.

Nike serves as an example of how managers can create an environment that effectively encourages innovation and leads to improved organizational performance. Since its founding, Nike always stayed at the forefront of the industry in innovation and creativity, producing lighter running shoes and eye-catching designs. More recently, however, Nike manages innovative processes by linking them to its bottom line. Balancing the creative side of the business with a commitment to the logistics side of things has fueled Nike's growth on a global scale. Studying markets and applying innovative marketing strategies led to Nike's position as an international powerhouse, buying up other apparel brands and expanding its portfolio globally.[13] Similarly, the CEO of an Atlanta construction company, Joseph A. Riedel, advises managers to understand their market before applying innovation.[14] These examples exemplify the importance of market knowledge and market-sensing activities when innovating. Without market knowledge, firms do not align with ideal conditions, and the effort of innovation may be misdirected or altogether wasted.

One way for managers to promote internal conditions ideal for marketing strategy innovation is through the use of innovation courses. In innovation courses, managers are given challenging tasks with the goal of forcing participants to focus on what is critically important without peripheral distractions. Innovation courses are said to be growing in availability, with many such courses offered at cutting-edge research universities. The courses are primarily geared toward managing individuals—whether employees, customers, other managers, or other possible organizational personnel—when innovation is the end goal. They stress the measurability of innovation as a resource, a key focus for firms investing in such training. Thus, training may help managers oversee challenging work environments by emphasizing innovative marketing strategies while focusing on performance.

Last, our study shows the value of profile deviation and covariation approaches to the study of marketing strategy, which are used surprisingly little in academic marketing research. Traditional approaches, such as moderated regression, typically would depict relationships among strategic, firm, or environmental variables as a series of interactions between marketing strategy and each condition.[15] The advantage of the approach used in this study is that it involves an integrated view of the conditions influencing the effectiveness of a given strategic orientation. By constructing the ideal firm profile for marketing strategy innovation, we were able to determine the optimal specific set of characteristics that lead to enhanced performance under certain firm and environmental conditions. In testing our profile construction, deviations from the ideal profile resulted in lower organizational performance, reiterating that a particular combination of conditions is associated with greater performance for those firms choosing to innovate their marketing strategies.

Recommendations

Based on our results, we offer four recommendations intended to help firms striving for innovative marketing strategies. First, managers are encouraged to **build autonomy within their organizations**. Our results support the importance of an autonomous structure for innovative marketing strategies to succeed. Granting autonomy to employees in the execution of their jobs involves creating a broad framework for their responsibilities and then delegating many of the performance details of their jobs to the employees themselves. Such an approach empowers the employees and generally results in a greater degree of employee commitment to the organization. Providing positive feedback for strong performance can also enhance employees' confidence and ensure that they stay fixed on a path that supports a company's strategies. An example of a firm that has embraced this approach is Zappos, the online retailer of shoes and apparel. Zappos's employees are allowed (and to a greater extent encouraged) to determine for themselves how to resolve issues for customers. Unlike many other firms that develop detailed step-by-step manuals for such activities, the company has no customer response manual.[16] Although considered unconventional when compared with formalized customer response policies of many companies, Zappos's approach seems to be working, as evidenced by continued sales growth.

Second, managers should strive to **create a challenging work environment** for their employees. Managers should avoid having employees engage in the drudgery of routinized, unchallenging tasks. Instead, encouraging employees to take on challenges that may at first appear to be above their abilities can help engender confidence for even bigger and more demanding tasks. Employees can also gain experience that prepares them for growth, and our results indicate that innovative marketing strategies may be more effective in such environments. Procter & Gamble's "Early Responsibility" program illustrates this approach. Through this program, new employees take on responsibilities that enable them to assess implications of their work for the overall operations of their particular division. The idea is to empower employees to fully comprehend their roles in the broader scheme of the company, thereby ultimately enabling more successful strategies to emerge.

Another way managers can create a challenging work environment is to encourage creativity in job tasks and reward employees for pursuing creativity.[17] Related research has found new product creativity is directly linked to financial performance. As one manager remarked:

Marketing creativity expands the number of customers we can touch. It greatly expands our image and our brand. People don't know we have so many other options for our customers and business, and I think that creativity in marketing helps us to further change the landscape of what other people think.

Third, managers should **develop and implement a systematic process for scanning the environment and disseminating relevant insights**. Environmental scanning involves systematically exploring a firm's environment to better understand the nature and pace of change, as well as to identify potential opportunities or challenges.[18] The development of a robust customer knowledge process is argued to not only allow quicker and more comprehensive understanding of buyer needs but also to generate insight into the potential trajectory of such needs as they evolve.[19] Relevant insights generated from customer knowledge might also suggest competitive threats. The results of our study indicate that a robust scanning process is critical for success in developing and implementing innovative marketing strategies.

DuPont's experience in the 1990s epitomizes the potential benefits of adopting a vigorous environmental scanning process. In the early 1990s, DuPont senior management noticed slowed growth across certain businesses—namely, Dacron polyester and nylon engineering resins. As sales declined, each of DuPont's businesses independently decided to shift its emphasis to more profitable market segments, most of which were at the higher end of the market. Competitors in the abandoned low-end segments ultimately capitalized on the increased volume they gained and were able to further reduce their costs and grow their market shares at DuPont's expense. A failure to generate and disseminate the relevant insights regarding customer preferences and competitive threats ultimately left DuPont unprepared in terms of its abilities to develop an effective marketing strategy. Learning from its failures, the firm developed processes for anticipating competitive threats early and for leveraging the insights generated to inform the development of preemptive strategies.[20]

Finally, whether related to customers or competitors, firms must **stay on top of market conditions**. Much like the prior recommendation of developing and implementing a systematic process for scanning the environment, firms must pay particular attention to market conditions. External to the firm are the micro and macro environment factors that influence the ability of the firm to operate as strategized (see Figure 11-2).

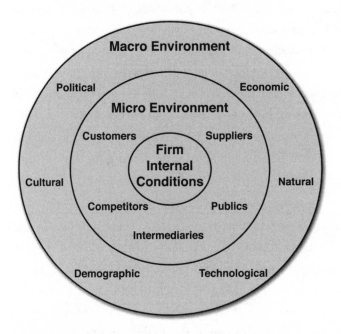

Figure 11-2 Micro and Macro Environmental Factors

Companies often face demographic shifts, new rivals, new technologies, new regulations, and other environmental changes that may influence their abilities to develop and successfully implement innovative marketing strategies, and our survey study results suggest that innovative strategies may be less risky when markets are more stable.

Thus, understanding the volatility in a firm's market may be critical for its ability to time innovative marketing strategy development.

The questions firms should use to examine market conditions often require an open-ended structure and result in inexact answers. Johnson & Johnson's strategy process exemplifies such an approach. In the early 2000s, in an effort to anticipate opportunities and position the company to develop appropriate strategies, the firm's executive committee and members of a strategy task force asked themselves four questions:

1. What will the demographics look like five to ten years in the future?
2. What will a typical doctor's office look like?
3. What role will governments play in the industry?
4. What role will payers play?

Traditional scanning activities may not uncover trends in market conditions. Alternative approaches that can be particularly effective in bringing such conditions to light include scenario planning to imagine various futures examining the actions of close rivals, or assessing why particular customer segments may have been lost. Attending conferences and events in other industries may also uncover market conditions not evident from conventional knowledge processes.[21]

Conclusion

The study reported here concludes that marketing strategy innovation is associated with enhanced performance under conditions of greater autonomy, challenging work environments, greater knowledge of the marketing environment, and a somewhat lesser degree of market turbulence. The results imply that managers may be better able to enhance performance by determining whether marketing strategy innovation should be prioritized depending on whether the conditions they face support such a strategy. In addition, managers who are seeking to develop and implement innovative marketing strategies as a way to compete may be more effective in implementing such innovative strategies to the extent the strategies are integrated with the firm's internal characteristics. Neglecting internal or external considerations can prove costly to innovation efforts and organizational performance. Such singular focus fails to consider the interrelated nature of the internal and external environment and risks misalignment with marketplace demands.

Endnotes

1. Kelly Hewett is an Associate Professor of Marketing, and Adam Hepworth is a doctoral student in Marketing, in the University of Tennessee's Haslam College of Business. Sharon Watson is an Associate Professor of Management at the University of Delaware.

2. 3 Marketing Lessons from the "Share a Coke" Campaign. May 7, 2015. Available at www.mayecreate.com/2015/05/3-marketing-lessons-from-the-share-a-coke-campaign/.

3. Andrews, J.L. and D.C. Smith. 1996. In Search of the Marketing Imagination: Factors Affecting the Creativity of Marketing Programs for Mature Products. *Journal of Marketing Research* 33(2): 174–187.

4. Favaro, K. 2013. Does P&G Need Product Innovation or Strategic Innovation? *Forbes*, July 12. Available at www.forbes.com/sites/boozandcompany/2013/07/12/does-pg-need-product-innovation-or-strategic-innovation/.

5. Dada, O., and A. Watson. 2010. Entrepreneurial Orientation and the Franchise System: Organisational Antecedents and Performance Outcomes. *European Journal of Marketing* 47(5/6): 790–812.

6. Van de Ven, A.H. and R. Drazin. 1985. The Concept of Fit in Contingency Theory. In *Research in Organizational Behavior*, ed. L.L. Cummings and B.M. Straw, 333–365. Greenwich, CT: JAI Press.

7. Chong, D. SF, et al. 2011. A Double-Edged Sword: The Effects of Challenge and Hindrance Time Pressure on New Product Development Teams. *Engineering Management, IEEE Transactions on Engineering Management* 58 (1): 71–86.

8. Podsakoff, N.P., J.A. LePine, and M.A. LePine. 2007. Differential Challenge Stressor–Hindrance Stressor Relationships with Job Attitudes, Turnover Intentions, Turnover, and Withdrawal Behavior: A Meta-analysis. *Journal of Applied Psychology* 92 (2): 438.

9. McCabe, L. 2015. Inbound for Everyone Everywhere: HubSpot's Ques. September 11. Available at www.it-director.com/blogs/Laurie_McCabe/2015/9/inbound-for-everyone-everywhere-hubspots-ques.html.

10. Leinbach-Reyhle, N. 2015. Prime Day Fail or Prime Day Success? Let the Numbers Be the Judge.... July 15. Available at www.forbes.com/sites/nicoleleinbachreyhle/2015/07/15/prime-day-fail-or-prime-day-success-let-these-numbers-be-the-judge/.

11. Lipshutz, J. 2014. U2 Releases Free Album 'Songs of Innocence' at Apple Event. September 9. Available at www.billboard.com/articles/news/6244130/u2-songs-of-innocence-free-album-apple-itunes.

12. Dibrell, C., S. Fairclough, and P.S. Davis. 2015. The Impact of External and Internal Entrainment on Firm Innovativeness: A Test of Moderation. *Journal of Business Research* 68(1): 19–26.

13. Holmes, S. and A.A. Bernstein. 2004. The New Nike. *Business Week*, Issue 3900, 78–86.

14. Donath, B. 2000. Business Must Come Before Creativity. *Marketing News* 33(6): 8–9.

15. Zajac, E.J., M.S. Kraatz, and R.K.F. Bresser. 2000. Modeling the Dynamics of Strategic Fit: A Normative Approach to Strategic Change. *Strategic Management Journal* 21(3): 429–453.

16. Ji-Eun, YE. 2012. 7 Principles for Employee Empowerment. *SERI Quarterly*, October, 5(4): 90.

17. Im, S., and J.P. Workman Jr. 2004. Market Orientation, Creativity, and New Product Performance in High-Technology Firms. *Journal of Marketing* (68)2: 114–132.

18. Conway, M. 2009. Environmental Scanning: What It Is, How to Do It.... *Thinking Futures*, April. Available at http://thinkingfutures.net/wp-content/uploads/2010/10/ES-Guide-April-09.pdf.

19. Jayachandran, S., K. Hewett, and P. Kaufman. 2004. Customer Response Capability in a Sense-and-Respond Era: The Role of Customer Knowledge Process. *Journal of the Academy of Marketing Science* 32(3), 219–233.

20. Day, G.S., and P.J. Schoemaker. 2005. Scanning the Periphery. *Harvard Business Review* 83(11): 135.

21. Schoemaker, P.J., S. Krupp, and S. Howland. 2013. Strategic Leadership: The Essential Skills. *Harvard Business Review* 91(1), 131–134.

12

External Barriers to Integration: Tearing Down the Walls

By Chad W. Autry, Stephanie Eckerd, and William J. Rose[1]

Once upon a time, there were two large and very complex companies. One, which we will call Company G, manufactured and distributed gifts, seasonal/holiday products, and home decorations. The other, which we will refer to as Company P, produced and sold packaged food and beverages, as well as personal care products. Company G and Company P historically had much in common. Both started out over a century ago as small family businesses based in North America, before eventually growing into multinational corporations that now have many divisions, products, employees, and customers. Both originally sought to serve mass consumer markets by distributing their manufactured goods through independent retail outlets. And both eventually captured a significant market share within their respective industries—even today, customer surveys indicate that people generally like, and in many cases love, both firms' products. However, as of 2015, one important difference between these two companies stands out: Company P greatly outperforms Company G in terms of operational coordination, cost efficiency, and order fulfillment capabilities, and as a result, P yields much greater shareholder value than G could seemingly dream of. In fact, while P is often cited as a global leader within its industry and is frequently included on "Top Ten" lists published within global periodicals due to its financial and operational performance, G's name is rarely found on the same lists. Given the multiple similarities and striking differences, one question lingers: how is such a disparity in operational and financial outcomes possible, given these two firms' similarity in heritage?

In order to understand why two companies that took such similar paths ended up at such different destinations, it is necessary to briefly examine their journeys. Both companies started out small and eventually grew large. When small firms, especially manufacturers, grow beyond a single entrepreneur's span of control, they tend to adopt a structure that includes multiple functionally focused divisions, and as they keep growing, more and more such divisions are created that each focus on a singular task or homogenous set of tasks. From an efficiency standpoint, such a structure makes intuitive sense: each workgroup focuses on what it does best, i.e., *specializes* in performing related tasks within a narrow scope of expertise that lead to a particular goal achievement or valued output. For example, the sales function is responsible for selling, the R&D group for developing new offerings, and so on. Under the strategic philosophy of functional specialization, the effectiveness of each functional area's output should be maximized, and as the theory goes, the sum of these effectiveness episodes is aggregated in the most efficient way possible. The sum of the parts should yield output that is both efficiently created and effective for satisfying customers—in most cases taking the form of a product or service—and should lead to the greatest possibility of financial success for the business as a whole.

However, the adoption of a philosophy of specialization as an underlying "operating system" requires that a certain condition can be assumed to be true: that the multiple functional areas of the business can be seamlessly sewn back together for the purpose of creating a value-laden output via the integration of business processes. Specifically, when businesses grow and functional specialization is adopted as a strategy, their parts effectively disintegrate by design, but in order for the business to perform optimally, these functional parts must be reintegrated such that the different functions act as a single, contiguous whole. As one manager we interviewed for this research put it, integration implies *that multiple, independently acting functional entities are taught to behave as though they are one.*

And this is where our two companies from the opening example differ. Company G, despite possessing some of the most sought-after products, innovative marketing programs, and desirable distribution arrangements in its industry, has traditionally struggled greatly to integrate its business functions into a single, cohesive business process having a unified mission, vision, and values. As a result, though the company is successful from a revenue standpoint, it has often failed to deliver much potential value to its ownership group. Alternatively, Company P has for a long time made supply chain integration a strategic priority from the C-suite downward (to the extent that it is popularly cited as one of the most integrated firms in business history), even though its products and services, corporate locations, and marketplaces are scattered around the globe. Company P is considered to be a blue-chip stock by market analysts, and has consistently delivered value to stakeholders at a level greater than the market as a whole. The differences in outcomes witnessed between these two firms is stark, and can be attributed to their differences in integration both internally (within the functions owned and operated by the business) and also externally to the firm itself (including functions such as advertisement and logistics, which are outsourced to other companies), and the contrast should thus serve as a lesson for other businesses' leaders to follow. But alas, unfortunately, many companies struggle to even define integration, and even those who do tend to fall short of actually achieving the gamut of benefits integration offers, for various reasons.

The purpose of this chapter is not to explain what integration actually is (which by now, if you've read the chapters that precede it, should be somewhat clear). Rather, our goal here is to explain what precisely it is that stops or hinders companies from integrating their processes across functions and/or firms in the first place. In short, this is a chapter about the barriers to integration. In it, we hope to illustrate the reasons that process integration across business functions (whether held and controlled internally or outsourced to partners) often proves very difficult for a company to achieve. More importantly, we conclude by describing some strategies that companies who are failing at integration can use to either mitigate the damage that comes with poorly connected processes, or improve operations such that additional value can be created.

Conceptualizing Integration

If you've read this far in the book, you know by now that there are multiple different perspectives on what integration actually is. However, the common thread among them seems to be that integration represents the capability of an organization to make its various activities and the functional groups they reside in work together as seamlessly as possible, whether the functions are actually managed by the firm or by one of its partners (such as a supplier, service provider, or customer). Much as a conductor coordinates different instrument groups within a symphony, the integration of a firm implies that all of the various functional groups are playing "in synchronization, and from the same sheet of music." In other words, integration means getting multiple different functional groups, existing within or across the formal boundaries of a company, to act in perfect synchronization toward the achievement of commonly understood and valued goals. However, based on our reading of the integration literature, there remain some gaps in our understanding.

First, when viewing different essays, publications, etc., there remains some question as to exactly "what" must be connected for integration to occur? Our best understanding of this issue leads us to believe that integration is a multidimensional concept inclusive of at least the following three aspects:

1. **Information channels**—For a firm's supply chain to be integrated, there must be "seamless" sharing of information between the functional entities that are working together to achieve its goals, consistent with its mission and vision. This means that the construction of information channels that allow for a free, unbiased, and multidirectional flow of data to occur is a necessary (but not sufficient) condition for supply chain integration to occur.

2. **Process ties**—For the firm's supply chain to be integrated, there must be a clearly defined and closely managed set of end-to-end business processes that all of the participants in the chain agree on as being critically important (at least in principle), which each function interfaces with and contributes to in light of the needs and requirements of the other functions, in order for value to be delivered to the customer at the end of the chain.

3. **Relational "glue"**—For the supply chain to be integrated, there needs to be a social basis for connecting information and processes across those functional entities. After all, unless a business is somehow fully automated, there are human beings working within these functional areas, and though people generally will do what's best for the company at large, nobody's perfect. People within the different areas may—and often do—act in ways that put personal gains or group preferences ahead of the outcomes of the company. In order for different functional groups to work together to achieve common goals, it helps if their members have a *social capital* basis to draw on, in the form of a functioning relationship.

Consideration of these three aspects leads us to a second conclusion pertaining to integration. Businesses, when well run, employ a set of commonly understood processes in order to achieve their desired output, but at the same time, these processes are embedded within social systems. Accordingly, there are both structural and social elements of informational, social, and relational connections that must be "tuned" with great coordination, in order for integration to be achieved. The failure to consider and act upon both social and structural elements of these aspects will almost certainly lead to a failure to integrate. These six theoretical dimensions must be addressed if companies are to enjoy the benefits of integration, and if mismanaged, they become barriers to integration that can be very difficult to overcome, and potentially damaging to the objectives of the business.

Defining the Types of Barriers

Based on our definition of integration, it is important to understand how the different types of barriers firms encounter might inhibit the achievement of integration. First, we suggest that integration barriers may be internal or external to the firm, and may be (to different degrees) controllable or uncontrollable. Regardless of their locality and controllability, they frustrate the positive outcomes associated with integration, and can make operational and financial performance elusive or even impossible to achieve. As such, our first step is to identify these barriers explicitly.

Information Barriers

Information sharing is core to the idea of integration. We see this reflected in many of the commonly adopted supply chain models—for example, the Supply Chain Operations Reference (SCOR) model and the Collaborative Planning, Forecasting, and Replenishing (CPFR) framework. However, information sharing also represents one of the more challenging areas to address, in part because there exist such strong structural barriers in addition to compelling social barriers against it.

Information-structural barriers. This subset of barriers consists of some of the more pressing obstacles preventing interorganizational (and sometimes even intraorganizational) information sharing. First and foremost among them involves the technologies required for organizations to expediently and reliably share information with one another. In considering these technology requirements alone, numerous issues arise. First, and perhaps most obvious, are the costs to implement commercial supply chain management software. A 2015 survey conducted by Software Advice revealed that small businesses, those defined as having revenues less than $50 million, expect to spend about $30,000 on software implementations this year; for medium- and large-size firms, that figure escalates to $171,000.[2] Complexity is another issue that quickly surfaces. The technologies available offer a dizzying array of options, and without adequate upfront planning, this can lead to additions, substitutions, and modifications along the way (which also serve to escalate costs even further). Planning in this regard is not as easily said as done, however, as even basic choices about whether to invest in off-the-shelf software that adheres to best practices versus those that are customizable involve numerous trade-offs and difficulty. The challenges detailed here have led to some rather spectacular and much publicized failures. Take, for example, Dell Computer and its attempt to transition to an enterprise resource planning (ERP) platform in the mid-1990s. After more than two years and an investment of over $200 million, Dell abandoned the project.[3] One of the noted issues was that the technology simply did not fit with the way it did business, and it was unlikely to achieve desired results from the effort.[4]

Information-social barriers. In addition to the structural barriers surrounding information exchange, there are also more intangible factors that frequently come into play. For many firms, the information that needs to be shared across supply chain partners in order to achieve integration is considered proprietary knowledge. This occurs commonly in situations where information needs to be supplied upstream, like product specifications to manufacturers or parts suppliers. Consider, for example, the now well-known case between Apple and Samsung. Apple has repeatedly accused Samsung of patent infringement, claiming the supplier appropriated the popular touchscreen technology that Apple invented for its own brand of Galaxy phones.[5] In other cases, the information required is financial data that firms feel compelled to keep under lock and key. As such, prohibitions against sharing these types of information with other organizations exist. This is a common concern for those sharing information downstream in the supply chain, and often for very good reason. For example, consider the supplier "squeezing" activity widely conducted by retailer Walmart and automotive manufacturer General Motors. These particular firms possess substantial purchasing power, and that makes it possible for them to demand to see suppliers' books for applying squeezing tactics. However, it also provides a clear illustration to suppliers regarding the risks associated with voluntarily sharing this type of data downstream.

Process Barriers

Process barriers exist because each firm has its own independent way of operating, processes that have developed over years or decades of operation, and that are ingrained both structurally and socially. Until the opportunity for integrating with another firm arises, this institutionalization of practices works fine. But once it comes time to merge independent processes, or make seamless the transition between them, it becomes apparent that wholesale process changes often will be required on one or both ends. And, not surprising, this change is hard. Who is going to change? How drastic are the changes? Is everyone on board? These are the kinds of barriers that must be tackled under the process domain.

Process-structural barriers. Here we address basic infrastructural considerations, aside from the specific technological aspects discussed previously. Under this category, we include governance structures, the competitive strategies and goals of the independent organizations, measurement issues, and limited resources. In fact, many of these same issues, such as goal alignment, have been long-standing issues within organizations; this was considered by Peter Drucker to be one of the "great divide" issues facing management. As such, extending appropriate process structures across firms represents a monumental challenge. Adopting the correct performance measures is critical in addressing the challenge. Most metrics are too isolated in nature, which provides an incomplete picture of supply chain performance and a substantial barrier to integration. To keep from just pushing problems to other nodes in the chain, holistic measures taken throughout the supply chain are necessary. This has been a criticism of many implementations of just-in-time (JIT) programs—that it resulted in lower inventories for one organization in the supply chain but just moved those inventories to other locations within the same chain.

Process-social barriers. The requirement for structural process adaptations leads to some change management issues on the social front, as well. Some of this is simply due to the added complexity involved with integrating separately functioning organizations. However, some of the more difficult challenges evolve around power dynamics and justice considerations, as well as employee motivation to make the requisite changes. For example, in the face of large-scale process changes (like those imposed by ERP implementation), employees are very sensitive to what the changes are and how they take place. In fact, ample evidence suggests that employees who are uncomfortable with the new processes will circumvent them if possible.[6] Justice considerations can greatly impact the adoption of new processes both internally and externally. Procedural justice, for example, describes how the actions supporting integration are taking place. If a new IT system is installed to support the effort, then how involved were the users in designating process roles and responsibilities, and did they have a voice in reflecting the new way in which work would be done? Distributive justice describes the outcomes of the integration effort. Are changes distributed equitably among member firms? Do users understand the benefits of integration efforts? The last item is key, as research shows that there are many misconceptions regarding the rationale for undertaking supply integration.[7] Without a compelling reason instigating the change, an already difficult change management task can be escalated to impossible to accomplish.

Relational Barriers

The barriers previously addressed are typically the more salient and observable barriers in need of removal. However, as evidenced above, if the people doing the work are not invested and committed to prioritizing and facilitating the integration effort, then removal of those other barriers will not be sufficient. This is some ways tantamount to facilitating a culture of integration, which can be quite difficult where previously there was none. One of the best corollaries to this involves the automotive industry in America in the 1980s. The Japanese car manufacturers were gaining market share at an extremely rapid clip over their American counterparts; the Japanese models were lower priced and of better quality. Executives from the American companies were actually invited out to Toyota's facilities in Japan and spent a great deal of time observing operations there. These executives then dutifully returned to their own organizations and began replicating everything they saw during their intensive tours at Toyota. But when they tried to replicate what they saw (the tools), they were wildly unsuccessful. Why? Because what they missed was the underlying relational fabric that held the lean system together, the philosophies and culture that undergirded the whole system and ultimately made it successful. Comparably, the power of the relational barriers to supply chain integration should not be minimized. It is critical to the success of any integration effort.

Relational-structural barriers. From a pragmatic standpoint, relationships (and the lack thereof) introduce complexities between functional units that can be difficult to overcome. Generally, relational-structural barriers pertain to differences in formal agreements, or understanding about how the units should relate to one another, or differences in understanding or interpretation of how the functional units (should) relate to the broader organization. Most commonly these pertain to the ways the units perceive risks and benefits, and the lens through which the unit should interact with the firm, in terms of its commitment, long-term orientation, and/or reporting hierarchies. Misaligned risk sharing and reward structures (which are informed by power dynamics and justice at the individual level of analysis) can create disagreements and lack of cohesion between functional groups. Similarly, if one group sustains a long-term orientation to decision making, while the other seeks quick gains and payoffs, differences in prioritization of initiatives are almost inevitably the result. These differences in structural elements cause unnecessary friction between group members and place stress on business processes that both groups are responsible for managing or enacting.

Relational-social barriers. Barriers within this category include trust, relational norms, opportunism, and cultural aspects. Trust is a multidimensional concept, involving competency trust and integrity trust, both of which are critical to integration efforts. Competency trust speaks to the skills and ability of the partner organization, whereas integrity trust addresses alignment of the morals and value systems of the integrating firms. One of the key reasons trust is so critical in these efforts is due to the long-term nature of integration efforts. Similarly, relational norms ideally develop and strengthen over time, also to facilitate the long-term orientation of integration. Relational norms include solidarity (a sense of unity between partners), flexibility, and mutuality (which reflects interdependence and reciprocity in the relationship). Another key benefit to achieving high levels of trust and relational norms is that they tend to minimize attempts at opportunism within the integrated supply chain. Opportunism can take many different forms, but several of the common manifestations include free riding, hold-up, and leakage. Finally, culture differences represent yet another barrier within the relational social category. Cultural differences can arise at the organizational level and the national level. One good example of a clash of national cultures was depicted in a 2004 paper describing the entry of Japanese logistics subsidiaries into Europe.[8] Trust building, flexibility, and service quality hallmarked the Japanese service strategy, whereas European managers' predominant focus centered on short-term costs.

Summary of the Barriers

As explained in the preceding sections, six types of barriers to integration, based on three aspects of interaction and two levels of analysis, can inhibit firms from truly integrating their functionality. However, it is also extremely important to note that these barriers do not exist in isolation of one another. Clearly, there are very strong correlations between them. For example, the persistence of mistrust between the employees of a firm's purchasing function and its logistics function (a relational-social barrier to integration) can lead to subsequent information hoarding within the purchasing function during the execution of sourcing activities (an information-structural barrier to integration). As a consequence of these correlations, barriers can become recursive or contingent on one another such that "death spirals" of disintegration occur. Managers should be aware of "payback loops," retribution, political revenge possibilities, etc., that can result from simple and possibly innocuous initial breakdowns in integration between functional units. An ounce of prevention may indeed alleviate the need for a gallon of cure.

Additionally, we once again channel Peter Drucker, who proclaimed that "culture eats strategy for breakfast." Drucker's exclamation forces us to recognize that integration begins and ends at the top of the organization, where leaders enact cultural norms of interfunctional behavior and relations. When cultural differences or problems filter down into the organizational structure, they manifest as social barriers, which over time become formalized as business rules that spawn process barriers. For interfunctional integration to be realized, top management, beginning with the CEO and Chief Supply Chain Officer, must set incentives and create an internal climate that supports collaboration across functional silos, and when necessary, with external partners. And it's not enough to simply pay lip service or provide encouragement for "working together." The best way to ensure that functions that ostensibly have different micro-level goals will be comfortable in working together is to find ways to blend their employees' KPIs that encourage alignment, communication, partnerships, quantification of outcomes, and interdependence.

How Can the Barriers Be Overcome?

As can be seen from the previous section, numerous barriers to integration exist that create problems for modern businesses. However, we would be remiss if we failed to note that the existing academic literature has discussed several ways in which supply chain integration can be facilitated. In particular, research has supported that relationships (both between internal functions and across separate firms) that are aligned, communicative, structured, quantified, and interdependent are the best candidates for integration, and when these conditions don't exist, their creation is a critical success factor for achieving cross-functional integration. As a result, firms seeking to integrate should focus on the particular relationships that exhibit these five characteristics, or seek to build them.

Alignment focuses on the different roles and responsibility of partners in the supply chain. Aligned partners share common goals and operating procedures based on clear mission statements. A shared objective inspires firms to work together and facilitates integration.

Communicative supply chain relationships are those characterized by a high degree of information sharing. Sharing information
allows supply chain members to connect with one another, allowing
for more accurate forecasts and further enabling integration across the
supply chain. Additionally, emphasizing the importance of communication throughout the supply chain at the personal level also builds
trust between partners, another necessary prerequisite for integration. Aligned and communicative partners share the tools, processes,
and information necessary to improve supply chain integration.

Structured partnerships are built on a foundation of shared risks
and rewards and seek to make interaction between partners as efficient as possible. Developing guidelines for choosing and managing
key relationships allows firms to enact a common vision and invest
resources into the integrated supply chain. Building cross-functional
and cross-organizational teams and establishing positions devoted to
managing supply chain partnerships build integration into the firm
itself.

Along with committing to communication, common goals, and
shared risks and rewards, **quantified** relationships allow partners to
ensure everyone involved has done their part. Quantifiable metrics
help to clearly define expectations for both partners and customers.
Additionally, quantifying expectations and outcomes enables firms to
assign responsibilities and rewards as well as keep tabs on where every
partner stands within the integrated supply chain. Without measuring
partner inputs or outcomes, firms seeking integration remain in the
dark about the facilitators, barriers, and outcomes of the integration
effort.

Finally, **interdependent** partners can influence one another's
behavior through developing cross-functional processes and teams,
training employees to focus on the integrated supply chain instead
of their own functional silos, and implementing software to for-
mally connect different functions and firms. Building this interdependence early in the relationship through inclusion of partners in

cross-functional or cross-organizational meetings can help to solidify the partnership and improve communication throughout the integrated supply chain. Integration relies on multiple partners relying on one another for the shared benefits that come from supply chain integration.

Partnerships that reflect a high degree of alignment, communication, joint structures, measurable outcomes, and interdependence have a higher chance for successful supply chain integration. As a result, firms seeking integration should work not only to reduce barriers by training employees to develop strong key relationships and better assessing customer requirements, but also increase integration facilitators by aligning mission and goals and quantifying outcomes and expectations of one another. While researchers have examined the barriers and facilitators of supply chain integration, the literature offers little guidance on how to implement and maintain supply chain integration beyond simply reducing barriers and improving facilitators.

Concluding Thoughts

Our research suggests that six types of barriers exist that can greatly inhibit firms from "getting their act together" by integrating the functional groups within their internal and external supply chains. Yet, we are also able to identify five conditions under which integration appears to be increasingly possible. Returning to the example from the opening vignette: Company G has traditionally struggled to integrate its marketing, procurement, manufacturing, and sales functions because of both social and structural barriers. From a structural perspective, there are few incentives for these functions to collaborate, cooperate, or even communicate due to incentive structures that are completely function-specific—each group is measured and compensated under decision rules that are wholly agnostic to activities that are undertaken outside the function. Salespeople are evaluated purely on sales, and not on inventory, while manufacturing is evaluated purely on cost efficiencies, while completely disregarding the notion that customers don't typically purchase in huge lot sizes or economic order quantities. And so, in Company G, business as usual means optimization within the functions, while ignoring any notion of their interdependence. The challenge for supply chain leadership in a company such as this one is to create the alignment, communication, joint structures, measurable outcomes, and interdependence that make integration a possibility. Such changes would require much leadership, but then again, many fairy tales have a happy ending.

Endnotes

1. Chad W. Autry is the William J. Taylor Professor of Supply Chain Management, and Stephanie Eckerd is an Assistant Professor of Supply Chain Management, in the University of Tennessee's Haslam College of Business. William J. Rose is an alumnus of the Haslam College of Business and an Assistant Professor of Management at University College Dublin, Ireland.

2. Donati, M. 2015. SMEs Plan to Spend an Average of $30,000 on Supply Chain Software in 2015, Says Study. *Supply Management*.

3. Carroll, B.J. 2007. *Lean Performance ERP Project Management: Implementing the Virtual Lean Enterprise*, 2nd ed. Boca Raton, FL: CRC Press.

4. Buckhout, S., E. Frey, and J. Nemec Jr. 1999. Making ERP Succeed: Turning Fear into Promise. *Strategy + Business*, 15.

5. Tam, P-W., and M. Hytha. 2014. Apple, Samsung Agree to End Patent Suits Outside U.S. *Bloomberg Business*, online August 6.

6. Bendoly, E., and M.J. Cotteleer. 2008. Understanding Behavioral Sources of Process Variation Following Enterprise System Deployment. *Journal of Operations Management*, 26(1): 23–44.

7. Moberg, C.R., T.W. Speh, and T.L. Freese. 2003. SCM: Making the Vision a Reality. *Supply Chain Management Review*, September/October, 34–39.

8. Smagalla, D. 2004. Supply-Chain Culture Clash. *MIT Sloan Management Review*.

Index

Numbers

3TG, 187

A

aberrations, DSI (demand and supply integration), 160-163
  plan-driven forecasting, 161
Accenture, 219-220
accountability, 74
  DSI (demand and supply integration), 173
  ownership, returns management, 197
accounting, returns management, 183
Action Appliance, 179
action plans, IBCPS (interest-based collaborative problem solving), 43-44
active listening, 39
affinity exercise, 43
aha moments, 23
aligning
  operational execution with value focus, DSI (demand and supply integration), 74
  PLi, surveys, 106
alignment, 191
  lack of alignment, DSI (demand and supply integration) abberations, 163
  market and nonmarket strategies, Walmart, 140-141
  returns management, 193-198
alignment framework, 126-130
  company analysis, 128, 135-136
  competitor analysis, 128, 134
  country analysis, 128, 134-135
  industry analysis, 128, 130
    buyer power, 131-132
    competitive rivalry, 130-131
    new entrants, 133
    substitutes, 132-133
    supplier power, 132
  market strategy alignment, 128
  nonmarket strategy, 128
  stakeholder analysis, 128, 135

alignment of purchasing and logistics, 120
Amazon Prime Day, 270
  market turbulence, 282
apparel industry, demand and supply integration (DSI), 147-148
Apple
  barriers to integration, 304
  Songs of Innocence (U2), 283
approved provider transaction model, sourcing continuum, 213-215
Approved Supplier List (ASL), 216
Arrow Electronics, 251
asking back, 39
ASL (Approved Supplier List), 216
ATMI Inc., 251
automotive assembly, DSI (demand and supply integration), 55-57
automotive industry
  resource scarcity, 239
  Toyota, relational barriers, 306
autonomy, marketing strategy innovation, 276-277, 290

B

BAAS (business as a service), 87
balanced scorecards, 63
barriers to integration, 85-88, 302
  external barriers to integration, 297-300
  information barriers, 302-304
  overcoming, 309-311
  overview, 311-312
  process barriers, 304-306
  relational barriers, 306-308
  summary of, 308-309
basic provider model, sourcing continuum, 212-213
behavioral antecedents, 7
behaviors, 7
Bell Canada, 222-223
best practices
  best-in-class organizations, 91-92
  PLi, 107-108
    effective systems and processes that enable superior results, 116-118

313